Kapur & Suri's
BASIC HUMAN GENETICS

Kapur & Suri's
BASIC HUMAN GENETICS

Third Edition

Revised & Edited by

Dipali J Trivedi
MBBS MS (Anatomy)
Senior Associate Professor
Department of Anatomy
BJ Medical College
Ahmedabad, Gujarat, India

Paras S Shrimankar
MBBS MS (Anatomy)
Senior Associate Professor
Department of Anatomy
BJ Medical College
Ahmedabad, Gujarat, India

Foreword
TC Singel

JAYPEE *The Health Sciences Publisher*
New Delhi | London | Philadelphia | Panama

Jaypee Brothers Medical Publishers (P) Ltd.

Headquarters
Jaypee Brothers Medical Publishers (P) Ltd.
4838/24, Ansari Road, Daryaganj
New Delhi 110 002, India
Phone: +91-11-43574357
Fax: +91-11-43574314
E-mail: jaypee@jaypeebrothers.com

Overseas Offices

J.P. Medical Ltd.
83, Victoria Street, London
SW1H 0HW (UK)
Phone: +44-20 3170 8910
Fax: +44(0)20 3008 6180
E-mail: info@jpmedpub.com

Jaypee-Highlights Medical Publishers Inc.
City of Knowledge, Bld. 237, Clayton
Panama City, Panama
Phone: +1 507-301-0496
Fax: +1 507-301-0499
E-mail: cservice@jphmedical.com

Jaypee Medical Inc.
The Bourse
111, South Independence Mall East
Suite 835, Philadelphia
PA 19106, USA
Phone: +1 267-519-9789
E-mail: jpmed.us@gmail.com

Jaypee Brothers Medical Publishers (P) Ltd.
17/1-B, Babar Road, Block-B, Shaymali
Mohammadpur, Dhaka-1207
Bangladesh
Mobile: +08801912003485
E-mail: jaypeedhaka@gmail.com

Jaypee Brothers Medical Publishers (P) Ltd.
Bhotahity, Kathmandu, Nepal
Phone: +977-9741283608
E-mail: kathmandu@jaypeebrothers.com

Website: www.jaypeebrothers.com
Website: www.jaypeedigital.com

© 2016, Jaypee Brothers Medical Publishers

The views and opinions expressed in this book are solely those of the original contributor(s)/author(s) and do not necessarily represent those of editor(s) of the book.

All rights reserved. No part of this publication may be reproduced, stored or transmitted in any form or by any means, electronic, mechanical, photocopying, recording or otherwise, without the prior permission in writing of the publishers.

All brand names and product names used in this book are trade names, service marks, trademarks or registered trademarks of their respective owners. The publisher is not associated with any product or vendor mentioned in this book.

Medical knowledge and practice change constantly. This book is designed to provide accurate, authoritative information about the subject matter in question. However, readers are advised to check the most current information available on procedures included and check information from the manufacturer of each product to be administered, to verify the recommended dose, formula, method and duration of administration, adverse effects and contraindications. It is the responsibility of the practitioner to take all appropriate safety precautions. Neither the publisher nor the author(s)/editor(s) assume any liability for any injury and/or damage to persons or property arising from or related to use of material in this book.

This book is sold on the understanding that the publisher is not engaged in providing professional medical services. If such advice or services are required, the services of a competent medical professional should be sought.

Every effort has been made where necessary to contact holders of copyright to obtain permission to reproduce copyright material. If any have been inadvertently overlooked, the publisher will be pleased to make the necessary arrangements at the first opportunity.

Inquiries for bulk sales may be solicited at: jaypee@jaypeebrothers.com

Kapur & Suri's Basic Human Genetics

First Edition: 1984
Reprint: 1998
Second Edition: 1991
Reprint: 1995, 2005

Third Edition: **2016**

ISBN: 978-93-5250-027-7

Printed at Rajkamal Electric Press, Plot No. 2, Phase-IV, Kundli, Haryana.

Dedicated to

All students who inspired us
All the teachers of Anatomy who motivated us
and
Our parents and family members
who encouraged and supported us

Foreword

It is my pleasure and privilege to write a foreword for the third edition of *Kapur & Suri's Basic Human Genetics* written by senior teachers, Drs Dipali J Trivedi and Paras S Shrimankar. This book is very well structured and easy to understand. It introduces reader to the basic concepts of genetics. The chapters are nicely presented having necessary information with adequate figures and tables. Editors have made an excellent attempt to present the complex topics of subject in the most simplified way so that it can serve MBBS students, as well as the students from allied medical sciences such as Dental, Nursing, etc. The book is very well written and designed, so I hope, students would like to read and gain required knowledge of the subject.

TC Singel
Professor and Head
Department of Anatomy
BJ Medical College
Ahmedabad, Gujarat, India

Preface to the Third Edition

Genetics is a rapidly expanding field of the medical science. All components of the human body are influenced by the genes, so knowing the basic concepts of human genetics are essential in the diagnosis, management and prevention of various disorders. Every person related to medical field needs to know the fundamental aspects of the genetics. The third edition of *Kapur & Suri's Basic Human Genetics* is written after a long span of around 25 years. The book deals in initial chapters with introduction, structure of chromosome and its abnormalities. In this edition, efforts have been made to add information regarding a brief history of genetics and classification of chromosome. In the chapter of chromosomal abnormalities, attempt has been made to provide information related to more genetic syndromes. For better understanding of various chromosomal abnormalities, the text is augmented with related illustrations. The other chapters serve topics on genes, inheritance, population genetics, prevention and treatment of genetic diseases. The study of patterns of inheritance is important for the diagnosis, prognosis and estimation of the recurrence risk in other family members. To make it more understandable, different types of inheritance are presented with pedigree analysis. Emphasis has also been made on latest techniques for prenatal diagnosis. The present edition gives an overview of the *Human Genome Project*, which is the most ambitious project of biomedical researches.

It is hoped that the book will meet the requirements of every reader. We have tried to minimize the errors during preparation of the edition. We would be grateful to the readers for any feedback regarding lacunae in the content of the book.

Dipali J Trivedi **Paras S Shrimankar**

Preface to the First Edition

Basic Human Genetics is designed to be an introduction to the fundamentals of human genetics. The text has been arranged in small classified parts. It is adequately illustrated with simple line diagrams. Tabulated form of the text has been used for quick reading and lasting memory. The practical applications of genetics to clinical medicine are stressed throughout the book. The attempt has been made to meet the requirements of medical students. We shall be grateful to the readers for their suggestions to improve the book.

V Kapur
RK Suri

Acknowledgments

Our sincere thanks to Dr TC Singel (Professor and Head) and Dr CA Pensi (Ex-Professor and Head), Department of Anatomy, BJ Medical College, Ahmedabad, Gujarat, India, for their guidance and motivation. We would like to extend our gratitude to all senior teachers in the field of anatomy.

We are grateful to Dr Bharat J Shah (Dean/Principal), BJ Medical College for his encouragement.

We would also like to acknowledge our friends, colleagues and all our students who inspired and supported us.

We also wish to thank all our family members who have patiently accepted our long time preoccupation in this task. We are very happy to dedicate, *Kapur & Suri's Basic Human Genetics* to our family.

We gratefully acknowledge Shri Jitendar P Vij (Group Chairman), Mr Ankit Vij (Group President), Mr Tarun Duneja (Director-Publishing), Mr KK Raman (Production Manager) and all other staff of M/s Jaypee Brothers Medical Publishers (P) Ltd, New Delhi, India, for their role in executing the book successfully. We would like to thank Mr Sharad Patel (Commissioning Editor) for his help and cooperation throughout the process of preparation.

How can we forget the almighty God who gave us the insight and knowledge to accomplish this task.

Contents

1. **Introduction** 1
   - Genetics 1
2. **Chromosomes** 6
   - Karyotype 10
   - Fluorescent *in situ* Hybridization 15
   - Lyon's Hypothesis 18
3. **Cell Division** 21
   - Factors Influencing the Rate 21
   - The Cell Cycle 22
   - Mitosis 26
   - Meiosis 28
4. **Chromosomal Abnormalities** 35
   - On the Basis of Type of Abnormality 35
   - On the Basis of Type of Chromosome Involved 35
   - Numerical Abnormalities 40
   - Disorders Affecting Sex Chromosomes 48
   - Treatment 53
   - Intersex 54
5. **Genes** 56
   - Classification 56
   - Effects of Gene 57
   - Chemical Basis of Gene 58
   - Mutations 60
6. **Inheritance** 65
   - Inheritance 65
   - Mendel's Laws of Inheritance 66
   - Single Gene Inheritance 70
7. **Population Genetics** 82
   - Dermatoglyphics 82
   - Study of Twins 86
8. **Prevention and Treatment of Genetic Diseases** 90
   - Prevention of Genetic Diseases 90
   - Noninvasive Techniques 92

- Invasive Techniques *93*
- Treatment of Genetic Diseases *96*

9. The Human Genome Project and Recent Advances in Genetics   98
- Goals *98*
- Benefits of Human Genome Project *99*
- Recent Advances in Genetics *99*

Index   *103*

1
Introduction

Man's most precious treasure is his genetic heritage which guides the health and proper development of future generations. Undoubtedly, one of the most significant subdivisions of the medical science in the present-day world is Genetics, because all components of the human body are influenced by genes. Genetic disease is relevant to all medical specialities.

■ GENETICS

The term 'genetics' was coined by Bateson in 1906. It has been derived from the Greek word *'gene'* (gene = 'to become' or to 'grow into').

It is the study of science of the heredity. Heredity is transmission of similarities from parents to offsprings and the science also includes variations which often occurs in the offsprings.

Why to Study the Genetics?

Genetics, once largely confined to relatively rare conditions seen by only a few specialists, is now becoming a central component of our understanding of most major diseases. The rapid scientific progress in genetics has led to several practical implications for human well-being.

- It helps us to understand how normal variations between individuals are brought about.
- Knowledge of genetics is helpful in understanding the causation of diseases.
- A detailed study of the subject is important for understanding of disease process, prognosis and its effective management at the molecular level.
- Knowledge of genetics has also led to possible means of prevention of genetic disorders through genetic counseling and antenatal diagnosis.

- Genetics serves to solve even legal problems. Legal cases like disputed parentage and investigation of crimes may be sorted out by an analysis of blood groups, DNA fingerprinting or other inherited characteristics.

Genetics and Its Sub-divisions (Fig. 1.1)

The rapid development of human genetics during recent decades has created many interactions with other fields of medicine.

Cytogenetics

This branch of science deals with the study of chromosomes (normal and abnormal). Knowledge of structure of nucleus and its components (chromosome) has been of tremendous help in elucidating the physico-chemical aspects of heredity.

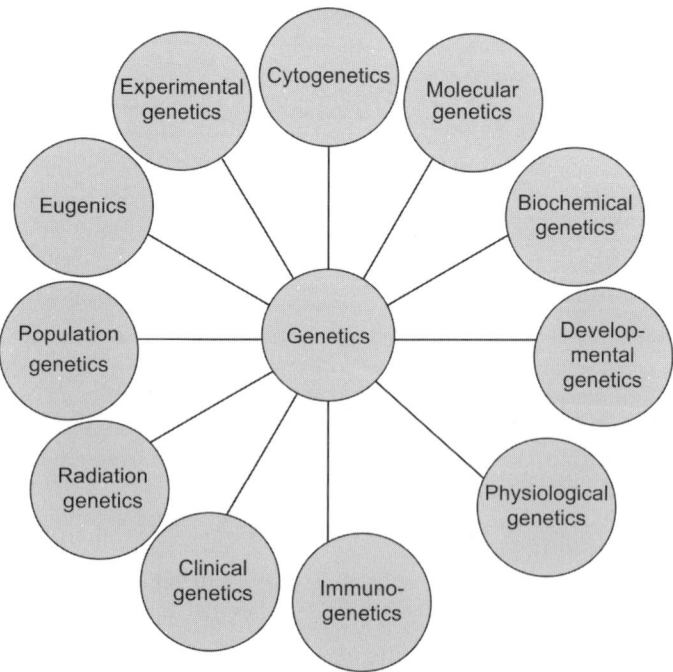

Fig. 1.1: Genetics in relation to other sciences

Molecular Genetics

It includes study of the molecular structure of the genetic material and its functional significance in the normal and abnormal state.

Biochemical Genetics

The science deals with the biochemical study of genetic material including the genetics of hemoglobin, inborn errors of metabolism etc.

Developmental Genetics

It is a branch of science which is concerned with the study of mechanisms involving genes during various stages of development.

Physiological Genetics

This branch of genetics involves the use of knowledge of physiology to elucidate the effects produced by genetic factors on an individual.

Immunogenetics

It is a branch of science which is concerned with the genetic aspects of immunity mechanisms.

Clinical Genetics

It is a branch of genetics which helps to establish the causative factors responsible for certain diseases, like diabetes mellitus, hemophilia etc.

Radiation Genetics

This branch of genetics involves the study of effects of various radiations on genes.

Population Genetics

It is the study of genes in populations. It deals with the distribution and behavior of genes and with how gene and genotype frequencies are maintained or changed in population.

Eugenics

It is a branch of science which deals with the application of principles of heredity for the improvement of mankind.

Experimental Genetics

It is the study of genetics by doing experiments on microorganisms and some laboratory animals. At the molecular level all the living organisms have same genetic mechanisms and therefore, the knowledge obtained from experiments on lower organisms can often be applied to man.

A Brief History

Before 6,000 years, stone carvings made by Babylonians illustrate the artificial cross-pollination of date palms.

Many years before the Christian era, Chinese farmers were improving varieties of rice by selecting the seeds from the plants that had the most desirable characteristics. Maize was cultivated and improved in the western hemisphere. Selection and hybridization were employed, although they were not even vaguely aware of the principles of genetics.

Hippocrates and Aristotle and other Greek philosophers made observations and suggested certain genetic principles. However, their results were vague and interspersed with errors. During this period, many tales regarding unusual hybrids were initiated, which perpetuated for about 2,000 years following the Greek period. They believed that giraffe was supposed to be a hybrid between the camel and the leopard. Banana trees were said to arise from hybridization between acacia and palm trees. These tales have contributed more towards the explanations of mechanism of reproduction rather than the principles of genetics.

O. Hertwig (1875) observed the entrance of sperm into the sea urchin egg. A single sperm was found to penetrate a single egg. This established a firm cytological basis of inheritance.

Gregor Mendel (1822–1884) (Fig. 1.2), an Austrian Monk considered as "Father of Genetics". He advanced in the field significantly by performing the series of experiments on the living organisms (garden peas). From the experiments, he formulated a series of fundamental principles of heredity and they were published in 1865. Unfortunately, his publications, which still formed the foundation of genetics, received little recognition for 35 years till it was rediscovered in 1900 by three scientists named Hugo de Vries, Carl Corrrens and Erich von.

Fig. 1.2: Johann Gregor Mendel

- Galton (1875) made a distinction between the effects of environment and heredity by the study of twins.
- Landsteiner (1900) discovered the ABO blood groups and initiated the genetics of blood group.
- Garrod (1902) reported for the first time alkaptonuria as an example of mendelian inheritance and initiated the concept of inborn error of metabolism.
- Tjio and Levan (1956), showed for the first time that the number of chromosomes in humans is 46 and not 48.
- Watson and Crick (1961) discovered the molecular structure of DNA.
- Hargovind Khorana and Marshall Nirenberg (1968) received the Nobel Prize for discovering the Genetic code. They determined the means by which a gene determines the sequence of amino acids in a protein.
- The Human Genome Project, started in 1990, provided the complete human DNA sequence in 2003. A Genome is a complete set of DNA, including all of its genes.

2

Chromosomes

Greek: Chromos = colored, Soma = body

■ INTRODUCTION

Chromosomes are the vehicles of heredity. During interphase of the cell cycle, they are coiled in the form of chromatin threads. However, during cell division, they become highly condensed and are then visible as dark distinct rod-like basophilic structures.

The term 'chromosome' was introduced into the scientific vocabulary by Waldeyer in 1888.

Number

The number of chromosomes in each cell is fixed for a particular species. In human beings, it is 46. This is called Diploid Number (2n). However, in spermatozoa and ova, the number of chromosomes is only half the diploid number, i.e. 23. This is called Haploid Number (n). These 46 chromosomes are arranged in the form of 23 pairs.

Types

One member of each pair of chromosomes is derived from the mother and the other from the father. Twenty-two out of these 23 pairs are identical in both the sexes and are known as Autosomes. The chromosomes in the remaining pair are called sex chromosomes. In females, both the sex chromosomes are identical and are called X-chromosomes. However, in males, the two sex chromosomes are not identical and are called X and Y chromosomes. Therefore, female germ cell always has X chromosome, whereas a male germ cell may either have an X or Y chromosome.

It is evident from Table 2.1 that if an ovum is fertilized by a sperm carrying X chromosome, it will result in a female child and, if fertilized by a sperm carrying Y chromosome, it will result in a male child. So, it is worth noticing that the sex of the child is determined by the father.

Table 2.1: Determination of sex

	Female Gamete	Male Gamete	Fertilized Ovum	Child
Sex chromosome	X +	Y	= XY	Male
Sex chromosome	X +	X	= XX	Female

Structure (Fig. 2.1)

Each chromosome is made up of two rod-shaped structures called chromatids. These chromatids are identical and lie parallel to each other. The two chromatids are united with each other at a pale-staining area called the centromere (Primary Constriction). The centromere divides each chromatid into two arms and is associated with the formation of spindles and chromosomal movements during cell division (Fig. 2.1). The free ends of the chromatids are known as telomere, which when intact, do not permit fusion with the adjacent chromosomes.

In certain chromosomes, another narrowing known as secondary constriction exists near one end of each chromatid. The part of chromatid beyond the secondary constriction looks like satellites. So, the chromosomes with such satellite bodies are known as SAT-chromosomes. These constrictions are said to be concerned with the formation of nucleoli. This is why they are also known as Nucleolar Organizers.

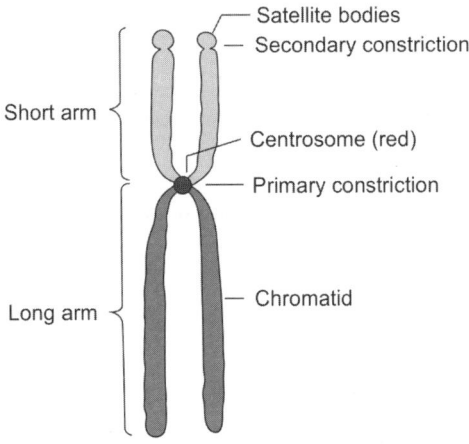

Fig. 2.1: Structure of a chromosome

Chromosomes may assume various shapes like twisted, spiral, curved or rod-like. Most of the chromosomes in human beings vary from 4 to 6 microns in length. They are the shortest during the metaphase of cell division.

Chemical Composition

The chemical constituents of chromosomes are:
- *Deoxyribonucleic acid (DNA)*: DNA is the most essential and stable molecular constituent of chromosomes. It is made up of deoxyribose sugar molecule and nucleotides. Each chromosome contains a single continuous double-stranded DNA molecule.
- *Ribose nucleic acid (RNA)*: Single-stranded structure having ribose as a sugar molecule.
- *Histones*: They are the basic proteins rich in arginine and lysine. They are aggregated along the DNA strand, which is coiled around each particle to form a complex body known as nucleosomes having 4 histones.
- *Acidic proteins*: They are nonhistone proteins and form many enzymes, e.g. DNA polymerase and RNA polymerase.

Significance

- Each cell of the body inherits from the fertilized ovum all the instructions necessary for proper organization and working of the various tissues and organs. Each chromosome bears on itself a large number of structures called genes; which guide the performance of particular cellular functions. This encyclopedia of information is stored within the chromosomes of each cell. Thus, each complete diploid set of chromosomes contain the cell's hereditary instruction or genome. The chromosomal threads bearing these instructions are known as Chromonemata or Genonemata.
- Cell activity is controlled by chromosomes which act by deciding the types of proteins synthesized within the cells.

Classification

Chromosomes can be classified on the basis of:
- Position of centromere
- Number of centromere
- Depending on function
- According to Denver system.

According to Position of Centromere (Fig 2.2)

- *Metacentric*: It is a chromosome with a centromere located in the middle of a chromosome. As a result of this, the two arms are almost equal.
- *Submetacentric*: It is a chromosome, with a centromere located slightly away from the mid-point. Consequently, the two arms are unequal. The longer arm is known as 'q' arm and the shorter arm is known as 'p' arm.
- *Acrocentric*: In this type of chromosomes, the centromere occupies subterminal position. One arm is very long and the other is short.
- *Telocentric*: It is a chromosome with a terminal centromere. Each chromatid, therefore, has one arm only.

According to Number of Centromere

- *Monocentric*: Having one centromere only, which is usual and normal.

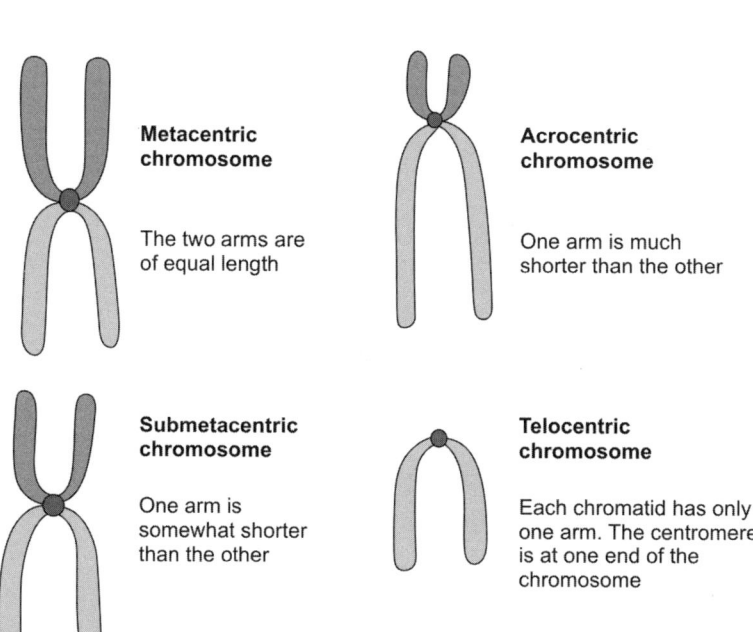

Fig. 2.2: Types of chromosomes

- *Dicentric*: Having two centromeres, which is found in some species of wheat.
- *Polycentric*: Having more than two centromeres seen in some forms of roundworms.
- *A-centric*: It represents only a fragment of chromosome having no centromere. It is not viable.

Depending on Function
- *Autosomes*: In man, there are 22 pairs of autosomes which are responsible for the determination of body parts and their functions.
- *Sex chromosomes*: There is one pair of sex chromosome in each sex. In males, it is XY and in females, it is XX. The sex chromosomes are responsible for the determination of sexes and their functions.

According to Denver System
In 1960, at a conference of Geneticists at Denver, human chromosomes including sex chromosomes are arranged into seven groups (from I to VII) depending on the size of chromosomes, which is popularly known as Denver system. According to modern convention, chromosomes are subdivided into Group-A to Group-G depending on the position of the centromere. This grouping is known as Patau's modification of Denver method. (Details of each group will be dealt with karyotyping).

Methods of Study
- To study the complete chromosome complement of an individual.
    - Karyotype preparation
    - FISH (Fluorescent *in situ* hybridization)
- To study sex chromosome constitution of a person, the following methods are available:
    - Study of Sex Chromatin or Barr Body
    - Study of Fluorescent Bodies in Buccal Smear
    - Study of Drumsticks in Polymorphonuclear Leukocytes

To study the complete chromosome complement:

■ KARYOTYPE
Karyotype is a complete chromosome set of a somatic cell. It also refers to a photomicrograph of an individual's chromosomes arranged in a standard manner.

Karyotyping

It is a process by which a karyotype is obtained.

Specimen

Chromosomes can be studied from the different tissues of the body. The basic principle involving cytogenetic preparations remain same for all tissues with slight modifications.
- Peripheral blood—most commonly used
- Skin fibroblast
- Bone marrow
- Chorionic villi ⎫
- Amniotic fluid cells ⎬ For prenatal diagnosis
- Fetal blood. ⎭

Chromosome analysis requires the provision of a large number of cells which are actively dividing. This is because it is only during critical stages of mitotic or meiotic cycle that chromosomes are in a suitable state to study.

Procedure (Fig. 2.3)

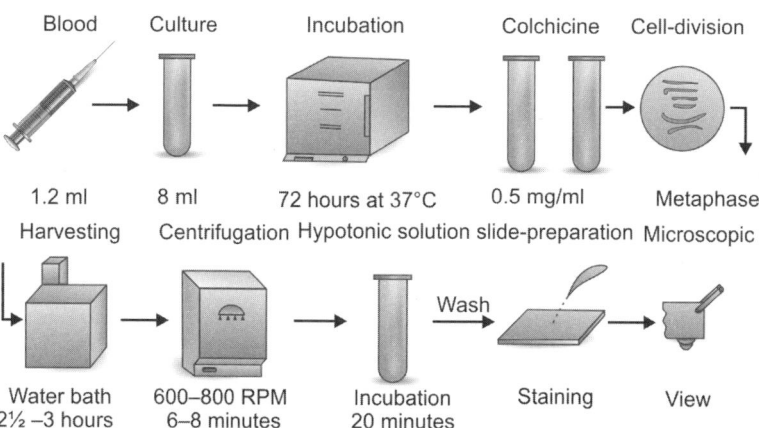

Fig. 2.3 : Procedure for karyotyping

Sample Collection
- Blood is collected from the peripheral vein under sterile conditions in Heparinized Vaccute.
- Planting should be done preferably within 4–5 hours of collection.

Planting

Under sterile conditions, blood sample is transferred to the culture tube containing the following constituents:
- *Culture medium*: Commonly used media are RPMI, TC 199, etc.
- *Neonatal calf serum/fetal bovine serum*: Added to nourish the culture cells.
- *Phytohemagglutinin*: A mitotic agent, added to increase the rate of mitosis of the cultured cells.
- *Antibiotics*: Penicillin and streptomycin combination is usually added to prevent the bacterial growth.

Incubation

The culture tube is kept in an incubator at 37° for 72 hours. During this period lymphocytes present in blood undergo mitosis.

Harvesting

- Around 69–70 hours after planting, colchicine is added to the culture tube to arrest the mitosis at metaphase by preventing the formation of spindle tubules.
- After 2 hours, cells are collected by centrifugation and then they are treated with hypotonic solution (0.56% KCL) so that the cells swell and chromosomes are dispersed.
- The hypotonic solution is discarded by centrifugation. Now the cells are treated with fixative solution containing acetic acid and methanol. Three such fixative washes are given to get cell suspension.

Slide Preparation

- Cell suspension is dropped from a height on chilled slides.
- Slides are allowed to dry at room temperature or on spirit lamp flame.

Staining

- For chromosome analysis, various banding techniques available are G-banding, Q-banding, R-banding, C-banding and NOR-banding. Commonly G-banding is practiced for study.
- *G-banding*: Slides with chromosome preparation are first treated with solution of trypsin. Trypsin denatures the chromosome protein.

Slides are then treated with Giemsa solution. The chromosomes show dark and light bands which can be observed under a microscope.

Microscopy

The stained slides are seen under the microscope. Good chromosome plates are identified and photographs are taken with the help of camera attached to microscope.

Preparation of Karyotype

It can be done as follows:
- Individual chromosomes are cut manually from the photograph, 22 autosomes including sex chromosomes are identified and arranged in pairs to prepare a Karyotype by using the following parameters:
  - Shape of the chromosome
  - Length of the chromosome
  - *Centromeric index:* This index is expressed in the form of ratio of the short arm length to the total chromosome length.

    $$\text{So, Centromeric Index} = \frac{\text{Short arm length}}{\text{Total chromosome length}}$$

    For example, in a metacentric chromosome, the centromeric index is 0.5.
  - *Proportion of the arms:* It is the ratio between the long and short arms of the chromosome. In a metacentric chromosome, this ratio is 1:1.
  - As per the location of bands along the length of chromosomes.

Recently, automatic karyotype system is available in which pairing of chromosomes is done automatically to prepare the karyotype of the patient (Fig. 2.4).

The 22 pairs of chromosomes can be arranged in 7 groups as per Patau's modification of Denver system. These are as follows:
- A group   : 1 to 3 pairs—Metacentric
- B group   : 4 to 5 pairs—Submetacentric
- C group   : 6 to 12 pairs—Submetacentric including X chromosome
- D group   : 13 to 15 pairs—Acrocentric
- E group   : 16 to 18 pairs—Submetacentric
- F group   : 19 to 20 pairs—Metacentric
- G group   : 21 to 22 pairs—Acrocentric including Y chromosome.

14   Basic Human Genetics

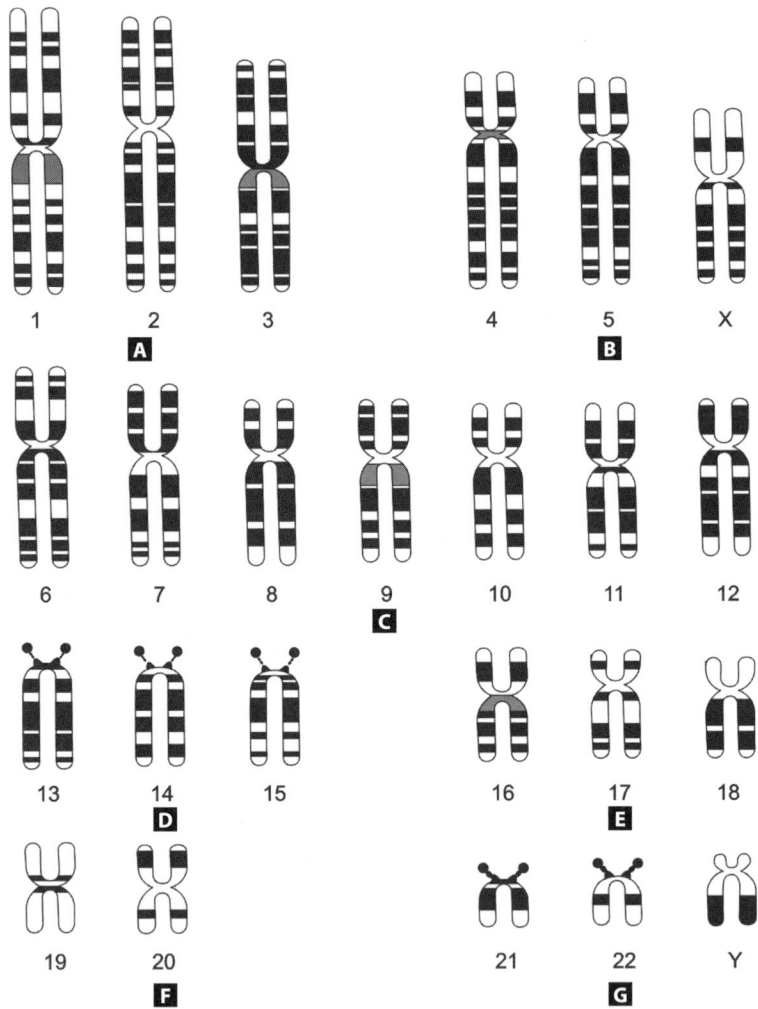

Figs 2.4A to G: Idiogram of a human male karyotype

Points to be Noted
- Groups A and F are metacentric
- Groups D and G are acrocentric
- Other groups are submetacentric

- In males, group G includes Y chromosome. Y chromosome is an acrocentric chromosome like others in the group but it shows two differences:
  1. It is usually the longest in group G.
  2. Its long arms are usually parallel to each other, which are divergent in the other members of group G.

The X chromosome is a member of group C and can be distinguished from other members of the group by banding techniques and by using special stains.

■ FLUORESCENT IN SITU HYBRIDIZATION (FISH) (FIG. 2.5)

We have seen that karyotyping is used to study the chromosomes of an individual. This helps to identify numerical as well as structural chromosomal abnormalities but it has limitations to identify very minute structural rearrangements or submicroscopic deletions.

FISH is based on the principle of DNA hybridization. In this technique, a labeled single-stranded DNA segment (probe) is exposed to denatured interphase or metaphase chromosomes. The probe undergoes complementary base pairing (hybridization) only with the complementary DNA sequence at a specific location on one of the denatured chromosomes. The site at which hybridization occurs on a particular chromosome can be visualized under a fluorescent microscope as the probe is labeled with fluorescent dye.

Fig. 2.5: Fluorescent *in situ* hybridization (FISH)

Uses of FISH

- To identify specific chromosome
- To identify missing or additional chromosomal material
- To know structural chromosomal defects especially microdeletions
- To assess the radiation effect or damage on a chromosome.

Types of FISH

Various types of chromosome-specific probes are available for FISH.
- *Centromeric probe*: Each chromosome has specific DNA sequences around the centromere. These probes are used to identify a particular chromosome.
- *Locus-specific probe*: The DNA sequences of homologous chromosomes and specific gene locations can be identified by locus specific probes. These probes are used in tumors, prenatal and postnatal samples.
- *Whole chromosome paint probe (WCP)*: Entire length of an individual chromosome is visualized by using this probe.
- *Multicolor probe*: Multiple probes labeled with different fluorescent dyes are used to paint all the chromosomes simultaneously. It helps to identify numerical chromosomal abnormalities.

To study the complete chromosome complement of an individual:
- Study of sex chromatin or Barr body
- Study of fluorescent bodies in buccal smear
- Study of drumsticks in polymorphonuclear leukocytes.

Sex Chromatin or Barr Body

Discovery: *Barr* and Bertram, in 1949, described these bodies in nuclei of phrenic nerve cells in a female cat. During interphase, somatic cell of a normal female presents a heterochromatin planoconvex body beneath the nuclear membrane. This is known as sex chromatin or Barr body. Out of the two X chromosomes in a normal female, one of them is highly coiled and the other member highly uncoiled. The highly coiled genetically inactive X chromosome forms the Barr body. These bodies help in nuclear sexing of the tissues.

Method of study: The cell nuclei of all tissues in a human female contain a sex chromatin body or Barr body (Fig. 2.6). But for convenience, the most suitable cells for study are those of buccal mucosa. The inside of cheek is gently scraped with a spatula and the cells obtained are spread on to a glass slide. This is called buccal smear. These cells are then fixed

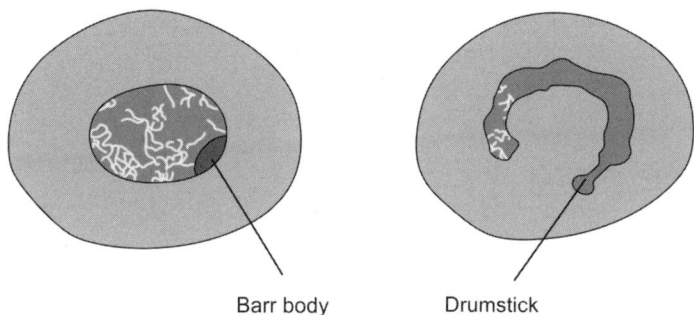

Fig. 2.6: Two different cells from a female showing Barr body in a squamous cell and a drumstick in a polymorphonuclear cell

and stained, after which they can be examined for the presence of Barr bodies. These bodies are observed in 30 to 60% of nuclei of a normal female.

Time of appearance: It appears in body cells when a female embryo is 2 weeks old.

Number (Fig. 2.7)

The number of Barr bodies in a cell is equal to the total number of X chromosomes minus one.
- Normal female (XX)—One Barr body
- Normal male (XY)—No Barr body
- Klinefelter's syndrome (XXXY)—2 Barr bodies
- Turner's syndrome (XO)—No Barr body
- Triple X syndrome (XXX)—2 Barr bodies.

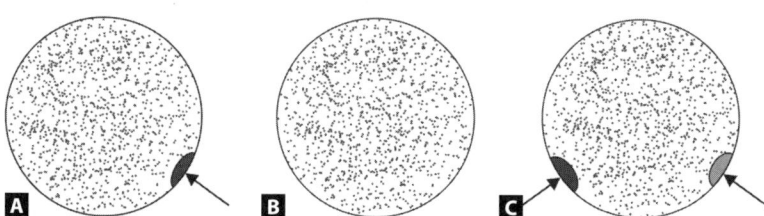

Figs 2.7A to C: Nucleus under different conditions; (A) Nucleus of a normal female (XX) showing single Barr body (marked by arrow); (B) Nucleus of a normal male (XY) does not show Barr body; (C) Nucleus of an individual with 2X chromosomes (e.g., Klinefelter's syndrome) showing 2 Barr Bodies (marked by arrow)

Significance: Presence of Barr bodies proves that the somatic cells of that individual contain 2 or more X chromosomes.

■ LYON'S HYPOTHESIS

This hypothesis was formulated by MF Lyon.

The hypothesis states that in the somatic cells of the female mammals, only one X chromosome is active and controls the formation of important enzymes and the other X chromosome is condensed and inactive. The inactive X chromosome appears in interphase cell as the Barr body. On the other hand, in males, there is only one X chromosome which is extended and carries out the normal body functions throughout the life. So, Barr body is absent.

Inactivation of X chromosome:
- Inactivation occurs in early embryonic life.
- It is random, i.e. the inactive X can either be paternal one or maternal one in different cells of the same individual.
- It is fixed, i.e. once the decision has been made regarding which X would be inactivated in a particular cell. All the descendants of that cell will show the same inactive X.

Significance of X inactivation:
- Dosage compensation
- Variability of expression in heterozygous females
- Mosaicism.

Dosage Compensation

The gene responsible for the production of G6PD enzyme is located on X chromosome. Though there are two X chromosomes in females and only one X chromosome in males, the amount of enzyme is equal in both the sexes.

Variability of Expression in Heterozygous Females

Variation in expression of X-linked disorders can range from completely normal to full expression of the abnormality. In heterozygous females, abnormalities are fully expressed when the X chromosome having abnormal gene is functional in a majority of the cells. Alternatively, she may be normal when X chromosome having abnormal gene is inactivated to form Barr body in majority of the cells.

Mosaicism

Women show mosaicism at cellular level, having two cell lines: one cell line with one X active, the other with the alternative X active.

Study of Fluorescent Bodies in Buccal Smear

In males, interphase nuclei of cells show a fluorescent spot called F-body or Y-chromatin when stained with fluorescent dye and examined under fluorescent microscope. The number of F-bodies indicate the number of Y chromosomes.

This technique can be used with buccal smears and so provides a method for assessing the number of Y chromosomes. It is the Y chromosome that determines the male sex and is responsible for the development of testes.

Since this technique is costly and the slide quickly deteriorates, it is usually not employed to study the sex chromatin status.

- Normal male (XY)—one F-body
- Normal female (XX)—absent F-body
- Turner's syndrome (XO)—absent F-body
- Klinefelter's syndrome (XXXY)—one F-body.

Study of Drumsticks in Polymorphonuclear Leukocytes

In suitably stained smears of peripheral blood, about 3% of polymorphonuclear leukocytes of females present a small accessory nuclear lobule, resembling a drumstick. It is not seen in polymorphs from males with XY and females with an XO sex chromosome constitution. The number of drumsticks does not bear any relationship to the number of X-chromosomes.

Importance of Chromosomal Studies

Chromosomal studies are useful:
- In the diagnosis of various chromosomal abnormalities like Turner's syndrome, Down's syndrome, Klinefelter's syndrome, etc.
- Clinically, in investigation of patients with abnormalities of sexual development or infertility.
- In the determination of sex of an unborn child.
- In large scale population surveys, e.g. to detect the effects of occupational hazards on chromosomes in relation to various environmental factors, like cold, heat, chemicals, etc.

- In new fields involving separation of X or Y bearing sperms.
- Chromosomal studies of chorionic villi and amniotic cells are helpful for hereditary disorders to know the risk of occurrence of the disease in the next generation.
- For identification of carrier status of a couple and to provide appropriate genetic counseling for the prognosis, management and recurrent risk estimation.
- Cytogenetic studies in malignancies, like chronic myeloid leukemia, help in prognosis and assessment of drug response.

3
Cell Division

■ INTRODUCTION
Multiplication of cells is carried out by division of pre-existing cells. Cell multiplication is of tremendous significance because of the following two reasons:
1. It constitutes an essential feature of embryonic development.
2. It is equally necessary for proper growth and for replacement of dead cells in the postnatal life.

Mechanism
Two distinct events take place in cell division:
1. *Karyokinesis:* It involves division of nucleus.
2. *Cytokinesis:* It involves division of cytoplasm.

■ FACTORS INFLUENCING THE RATE
- *Site*: The rate of cell division varies to a considerable extent in different tissues. The division of stem cells may be rapid in certain epithelia to replace the damaged cells as a result of mechanical factors, e.g. intestinal epithelium.
- *Demand*: The rate of cell division may also be influenced by the demand in a particular tissue. For example, it may rise to a peak during healing of wounded skin. So, the rate of cell division is very well coordinated with the demand for growth and replacement. Wherever this coordination is faulty, the tissues either fail to grow properly or they may overgrow, resulting in neoplasms.
- *Time*: There can be diurnal variation also in relation to cell division. For example, cell division in epidermis generally occurs at night (diurnal mitotic rhythm). Suggested explanation for this is based on the concept of chalones. During daytime, there is sufficient epinephrine in blood to form a complex with epidermal chalone and this complex inhibits cell division in epidermis.

- *Radiations*: Mitosis is inhibited on exposure to ionizing radiations. A typical feature of radiation sickness is the failure of epithelia to regenerate with consequent ulceration of skin and mucous membranes. These radiations may also result in failure of disjunction of chromatids.
- *Chemical agents*: Mitosis can be inhibited by many chemical agents like colchicine, vinblastine, etc. These agents inhibit the formation of spindle microtubules, thereby arresting the cell division in metaphase. This property is of great significance in cytogenetic studies, because chromosomes can be best studied in metaphase.

Normal Control

The exact mechanism of normal control of cell division is poorly understood. However, some of the observations in this regard are mentioned below:
- During early periods of embryonic life, there seems to be local control involving diffusion of metabolites among different cell groups, either stimulating or inhibiting the cell division.
- At later stages, general hormonal control also takes over. For example, thyroid hormones affect general metabolic rate, whereas corticosteroids and STH (Somatotropic Hormone) influence protein synthesis.
- In adult tissues, local control of cell division constitutes a significant factor in wound healing. A class of compounds, called Chalones, has been isolated from the normal tissue cells. These chalones are proteins of molecular weight ranging between 30,000 and 50,000. Normal cells are believed to produce chalones which exert inhibitory influence on cell division in their locality. In case of damage of cells, the concentration of chalones falls, which allows cell division to occur. As soon as the normal levels of chalones are restored, the division is again inhibited. These chalones are cell-specific but not species-specific.

■ THE CELL CYCLE (FIG. 3.1)

Most of the cells of the body have limited span of functional activity, at the end of which they divide into two daughter cells. The daughter cells divide further after their activity is completed. The period during which the cell is actively dividing is the phase of *mitosis* and the period between two successive divisions is known as *interphase*. Thus, cells that undergo division regularly continue to pass consecutively through interphase, mitosis, and so on.

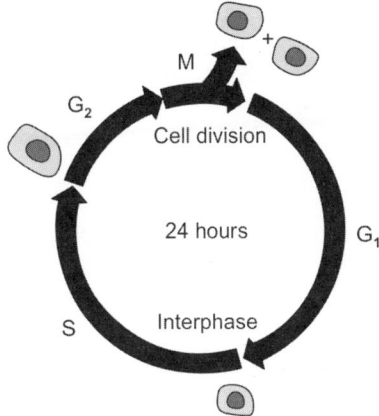

Fig. 3.1: Stages in cell cycle

One complete passage through the interphase and mitosis is termed as *cell cycle*. (Fig. 3.1)

Thus, the cell cycle is divisible into two main parts:
1. Interphase
2. Mitosis

Interphase

The cell carries out its ordinary work when it is in the interphase. Any cell that is not in the process of division is referred to as interphase cell.

During interphase, tightly coiled threads of chromosomes become uncoiled at certain sites. So, in an interphase nucleus, thread-like chromosomes have both coiled and uncoiled segments along their lengths. The uncoiled portions cannot be seen in an ordinary stained section with the light microscope, but coiled portions of chromosomes are stained and are visible as chromatin granules. As an interphase cell enters the process of mitosis, its chromosomes once again become coiled to look like rod-shaped bodies.

The interphase is sub-divided in G1 phase, S-phase and G2 phase.
- *G1 or preduplication phase*: Cells enter G1 phase after mitosis. Most of the cells of the body exist in this stage and carry out their functions. This phase covers the longest phase of the cell cycle. It may last as long as the life time.
- *S or synthesis phase*: It is the stage of DNA duplication. It lasts for about 7-8 hours.
- *G2 or postduplication phase*: It starts 5 hours before mitosis. This phase is utilized for the synthesis of proteins required for cell division.

Go phase
- Some cells do not undergo mitosis, e.g. neurons and cardiac muscle cells. These cells are said to be in *Go phase.*

Chromatin

These are particles of blue staining material in the interphase nucleus of a cell.

Types: It is of two types:
1. Extended chromatin
2. Condensed chromatin.

The two types of chromatins are compared in Table 3.1.

Table 3.1: Comparison of extended and condensed chromatins

Extended chromatin (Euchromatin) (Fig. 3.2)	*Condensed chromatin (Heterochromatin) (Fig. 3.2)*
1. It occupies mainly the central part of the nucleus	It occupies mainly the peripheral part of the nucleus It can be seen at 3 places: i. Close to inner surface of nuclear membrane ii. As clumps, scattered in the nuclear sap iii. Nucleolus-associated chromatin
2. It is dispersed and uncoiled. It stains poorly and is not visible under light microscope	It is clumped and coiled. It stains densely and is visible under light microscope as chromatin particles
3. Nuclei containing this chromatin are large and pale (since the amount of nuclear sap is directly proportional to the amount of extended chromatin). Such nuclei are called open face nuclei, e.g. nerve cells, hepatocytes	Nuclei containing this chromatin are small and dark, e.g. lymphocytes, cells lining the blood passages of liver
4. It is genetically active	It is genetically inactive
5. It contains preponderance of guanine and cytosine bases	It contains preponderance of adenine and thymine bases
6. It is more susceptible to the action of mutagenic agents	It is less susceptible to the action of mutagenic agents

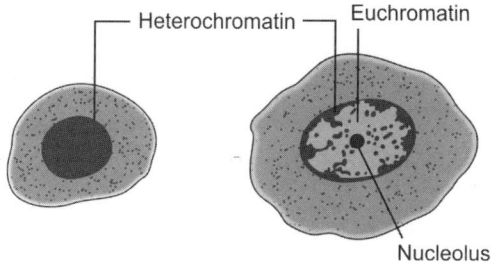

Fig. 3.2: Euchromatic and heterochromatic nuclei

There are evidences to indicate that the genes locked up in condensed chromatin during the interphase can become functional and provide instructions for new protein synthesis, if the need arises. One of the two X chromosomes in females is genetically inactive and contains condensed chromatin, seen as Barr body. Thus, females are in an advantageous position in having one extra chromosome as reserve in the form of Barr body. This accounts for more incidence of male embryos being aborted than females. Also, the females have more resistance to various diseases and ailments. It would not be wrong to comment that female sex, although supposed to be weak physically, is genetically strong.

Types of Cell Divisions

There are two types of cell divisions:
1. Mitosis
2. Meiosis

Mitosis is a fundamental process for life. During mitosis, a cell duplicates all of its contents, including its chromosomes and splits to form two identical daughter cells. Because this process is so critical, the steps of mitosis are carefully controlled by a number of genes. When mitosis is not regulated correctly, health problems such as cancer can result.

The other type of cell division, meiosis, ensures that humans have the same number of chromosomes in each generation. It is a two-step process that reduces the chromosome number by half, from 46 to 23 to form sperm and egg cells. When the sperm and egg cells unite at fertilization, each contributes 23 chromosomes. So, the resulting embryo will have the usual 46. Meiosis also allows genetic variation through a process of DNA shuffling while the cells are dividing.

■ MITOSIS

Mitos = Thread
Osis = Condition

Introduction

Mitosis or somatic cell division is the method by which body cells reduplicate themselves for the maintenance and growth of various tissues.

History

It was first observed in animal cells by Flemming in 1882.

Stages (Fig. 3.3)

Mitosis is a continuous process. It can, however, be divided into various stages for the convenience of description. These stages are:
- Prophase
- Metaphase
- Anaphase
- Telophase

Prophase (Prophase: Beginning; Before)

It involves the following changes:
- Chromosomes start coiling
- Each chromosome consists of two chromatids joined at the centromere
- Nucleolus and nuclear membrane disappearing
- Centrioles start moving towards the opposite poles. They produce number of microtubules that pass from one centriole to other and form a spindle
- Tubules radiating from each centriole create a star-like appearance or aster. The spindle and the two asters collectively form diaster.

Metaphase (Metaphase: Beyond; After)

- Centromeres of all the chromosomes come to lie in the equatorial plane
- Chromosomes become attached by the centromeres to the spindle microtubules

Cell Division 27

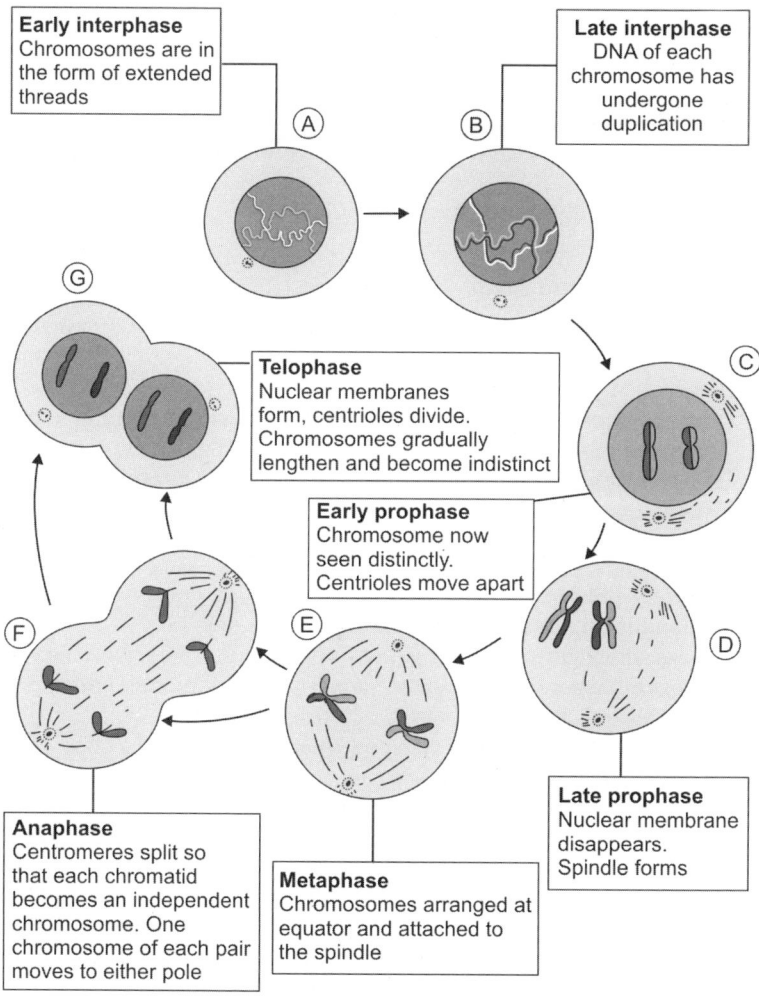

Fig. 3.3: Stages of mitosis

- Chromosomes get condensed and thus shortened. That is why, karyotype (the display of human chromosomes for analysis) is taken from metaphase as chromosomes are best visualized in this stage.

Anaphase (Anaphase: Again; Upward)
- Centromere of each chromosome splits longitudinally into two, so that the paired chromatids disjoin and become two new chromosomes
- The two new chromosomes move apart, one toward each pole of the cell
- Cytoplasmic division begins by infolding of the cell at the equator.

Telophase (Telophase—End)
- A constriction at equatorial plane deepens to form two daughter cells
- Nucleoli and nuclear membrane reappear
- The chromosomes elongate and become indistinct, forming chromatin network
- The centriole is duplicated at this stage or in early interphase.

Mitotic Figures

Cells in any phase of mitosis seen in sections are commonly said to contain mitotic figures. These mitotic figures are seen in:
- Places where growth is taking place
- In regions where maintenance of cell population is essential, e.g. bone marrow, lining epithelium of intestine
- Sites where repair process is in progress, e.g. fracture of a bone
- Abnormal cellular growths, e.g. cancer.

■ MEIOSIS

It involves two successive divisions called first and second meiotic divisions. In the interphase prior to the first meiotic division, DNA content of chromosomes is replicated, resulting in tetraploid amount of DNA, the number of chromosomes being diploid.

During Meiosis I: The DNA is reduced to diploid amount in each resultant cell, although the chromosome number is halved to the haploid.

During Meiosis II: The DNA in each new cell formed is reduced to the haploid amount, the chromosome number remaining haploid.

Meiosis I and II each, like mitosis, can be divided into prophase, metaphase, anaphase and telophase.

Meiosis I (First Meiotic Division)

Prophase

The prophase of first meiotic division is prolonged and different than prophase of mitosis. It is divided into five stages (Fig. 3.4).
1. Leptotene
2. Zygotene
3. Pachytene
4. Diplotene
5. Diakinesis.

Leptotene: The chromosomes become visible as individual threads, attached to nuclear membrane at one end. Distinct beads (Chromomeres) can be made out throughout their length. Individual chromatids cannot be seen (Fig. 3.4A).

Zygotene: Two chromosomes of each pair come to lie side by side forming a bivalent (Fig. 3.4B). This pairing of homologous chromosomes is known as synapsis or conjugation. Electron microscopic studies have revealed that homologous chromosomes are found to be held together by a fibrillar band, the synaptinemal complex. This band occupies the space, about 100 nm wide between them.

Pachytene: Each chromosome thickens and is now seen to be formed of two chromatids, joined at the centromere. Each bivalent, therefore, consists of four chromatids and is now known as a tetrad (Fig. 3.4C). A highly significant event now takes place. Two chromatids, one from each bivalent, become partially coiled around each other, thereby crossing at several points. This is known as decussation or crossing over and it involves exchange of DNA. Probably, synaptinemal complex plays some role. The points where crossing over takes place are called chiasmata (Fig. 3.4D). Exchange of genetic material by crossing over adds immense variety to the ultimate genetic makeup of a given individual.

Diplotene: The two chromosomes of a bivalent try to move apart. In this process, the chromatids get detached at the points of crossing over and the broken pieces are exchanged between the chromatids. Thus, exchange of DNA takes place. By this exchange of genetic material, there is reassortment of genes between maternal and paternal chromosomes. Each of the four chromatids of the tetrad now has a distinctive genetic content (Fig. 3.4E).

Diakinesis: The remaining chiasmata break and bivalent pairs move away from each other.

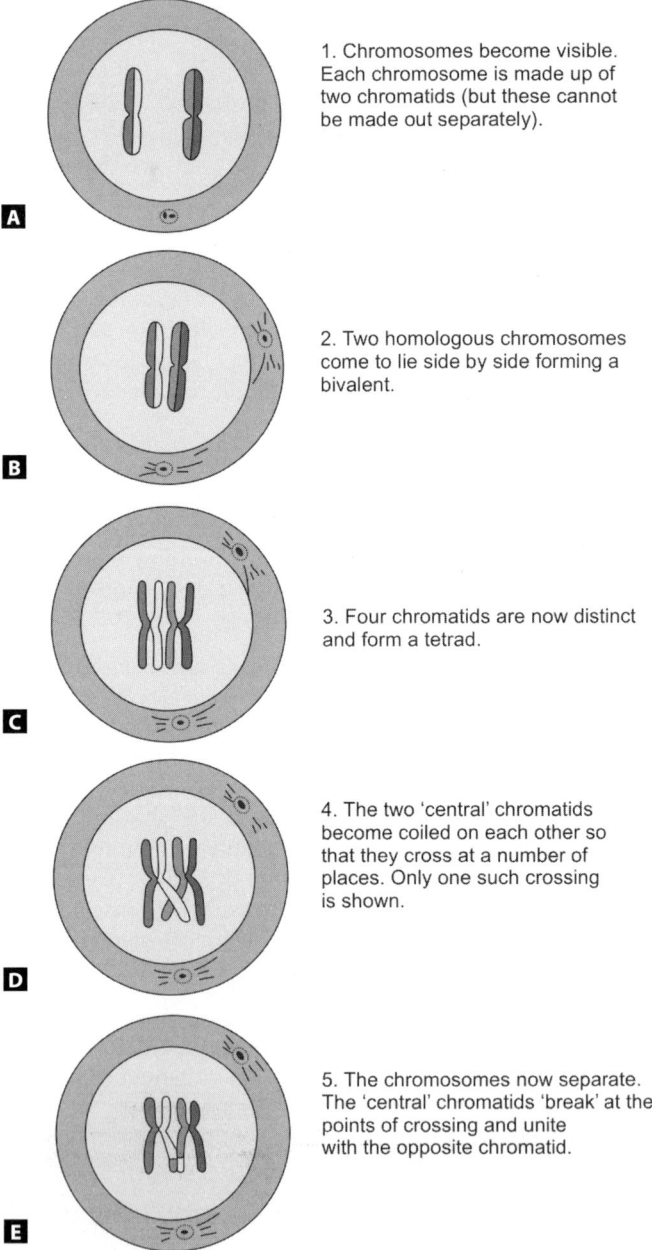

Figs 3.4A to E: Stages in prophase of 1st meiotic division

During the events of prophase, the nucleolus and nuclear membrane disappear.

Metaphase (Fig. 3.5)

It is similar to that of mitosis. The chromosomes get adherent to the spindle microtubules at the equator. The arrangement is such that the homologous pairs lie parallel to the equatorial plane, with one member on either side.

Anaphase (Fig. 3.5)

It is different from that of mitosis as the centromere does not split. Consequently, one chromosome of each homologous pair moves towards each pole of spindle. As a result, each daughter cell would, therefore, have 23 chromosomes (haploid number) each consisting of two chromatids.

Telophase (Fig. 3.5)

- It is similar to that of mitosis.
- Interphase preceding the second meiotic division.
- This interphase is unique, since no DNA synthesis occurs.

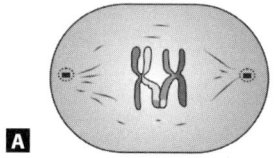

Metaphase
The nuclear membrane disappears. Spindle forms and chromosomes are attached to it by their centromeres.

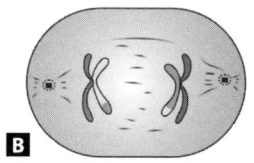

Anaphase
One entire chromosome of the pair moves to either pole. Note that the centromeres do not divide

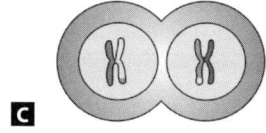

Telophase
Note that the chromosomes in each cell have been reduced to the haploid number

Figs 3.5A to C: Metaphase (A), anaphase (B), and telophase (C) of the first meiotic division

Meiosis II (Second Meiotic Division) (Fig. 3.6)

It is the same like mitosis, chromatids are separating during anaphase but unlike mitosis, separating chromatids are genetically dissimilar. Cytoplasmic division also proceeds and a total of four cells result from Meiosis I and II.

Significance of Meiosis (Fig. 3.7)

- It constitutes an essential event in the formation of gametes.
- It is important for restoring the chromosomal number, characteristic of species.
- It involves new combinations and exchanges of genetic material.

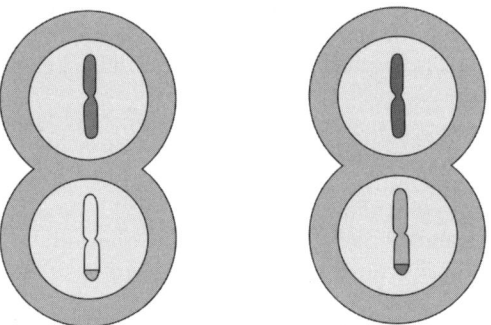

Fig. 3.6: Daughter cells resulting from the second meiotic division. The daughter cells are not alike because of the crossing over during the first division

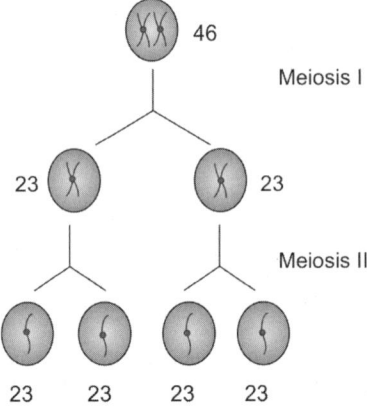

Fig. 3.7: Behavior of chromosomes during normal meiosis

The two types of cell division are compared in Table 3.2

Table 3.2: Comparison of mitosis and meiosis

Sr No	Feature	Mitosis	Meiosis
1.	Site	Somatic cells	Germ cells
2.	Number of divisions involved	Single	Two divisions I. Heterotypical (Reduction division) II. Homotypical (Equational division)
3.	Prophase	No crossing over, no chiasma formation	Prophase-I prolonged Crossing over + Chiasma + Exchange of genetic material +
4.	Chromosomal number	Remains the same	Reduced to half
5.	Chromosomal behavior	Independent of each other	Homologous chromosomes get paired together
6.	Daughter cell	Two daughter cells genetically identical	Four daughter cells genetically nonidentical

Abnormal Cell Divisions (Figs 3.8 and 3.9)

Most of the hazards of chromosomal numbers take place at anaphase. After splitting of the centromere, one or more chromosomes fail to migrate properly due to abnormal function of achromatic spindle. The phenomenon is known as *nondisjunction*. As a result, both the members of a particular pair go to one daughter cell which receives extra chromosome (trisomy), and the other daughter cell is deficient in that chromosome (monosomy).

Non-disjunction may take place in mitosis or meiosis, and it may involve sex-chromosomes as well as autosomes. Autosomal nondisjunction is less viable. Chances of survival is more with trisomic cells than monosomic cells. Turner's syndrome of female with 45, XO is probably the only example of viable monosomic individual.

When non-disjunction occurs in meiosis—I, all four gametes are abnormal (two with 24 chromosomes and two with 22 chromosomes).

When non-disjunction occurs in meiosis—II, two gametes are normal and two abnormal.

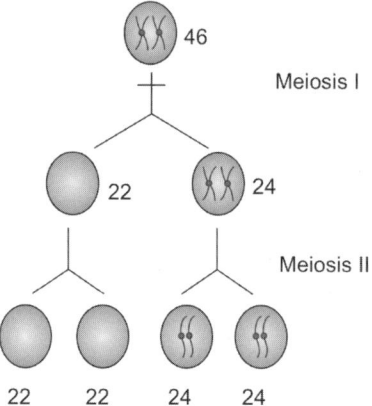

Fig. 3.8: Nondisjunction during first meiotic division

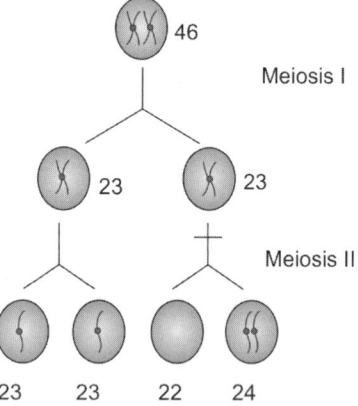

Fig. 3.9: Nondisjunction during second meiotic division

Nondisjunction in gametogenesis is observed in elderly females (40 years and above). Possibly the primary oocyte which starts first meiotic division in prenatal life completes the process just before ovulation after a prolonged interval of about 40 years or more. Delayed completion of first meiosis of oocyte might favour nondisjunction.

4
Chromosomal Abnormalities

Abnormal chromosomes are the vehicles of inherited abnormalities. Chromosomal abnormalities may be classified in a number of ways.

■ ON THE BASIS OF TYPE OF ABNORMALITY

- *Structural*: Involving change in structure of chromosomes. Structural aberrations result from single or multiple breaks along the chromosomal length. The broken fragments are then destroyed (deleted) or rearranged in various ways or shifted (translocated) to other chromosomes.
- *Numerical*: Involving change in number of chromosomes. Abnormality in number results almost invariably from the phenomenon of nondisjunction during the first and/or second meiotic division.

■ ON THE BASIS OF TYPE OF CHROMOSOME INVOLVED

- Involving autosomes, e.g. Down's syndrome
- Involving sex chromosomes, e.g.
    - Turner's syndrome
    - Klinefelter's syndrome.

Factors Responsible for Chromosomal Aberrations
- Ionizing radiation
- Viruses
- Chemical carcinogens
- Late maternal or paternal age
- Nondisjunction.

Structural Abnormalities

These abnormalities involve change in structure of chromosomes and may be categorized as follows:
- Deletion
- Ring chromosome
- Duplication
- Translocation
- Insertion
- Isochromosome
- Inversion.

Deletion or Deficiency (Fig. 4.1)

It is the loss of a portion of the chromosome. The clinical manifestations depend on the size of deleted portion and the function of the gene in that segment. Deletion may occur due to chromosome breakage within the chromosomes.

It is of two types:
1. Terminal
2. Interstitial

Terminal deletion: If there is a single break in the chromosome, only the terminal portion of the chromosome is deleted such deletion is known as terminal deletion, e.g. cri du chat syndrome 5p-.

It involves deletion of a portion of the short arm of chromosome number 5.

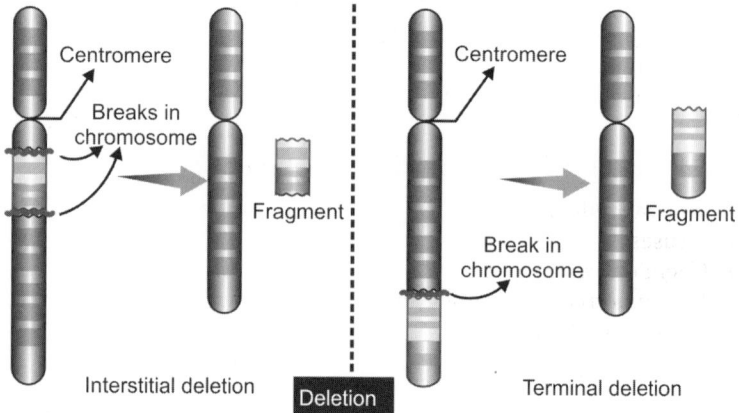

Fig. 4.1: Deletion

Interstitial deletion: If there are two breaks in a chromosome, the interstitial portion of chromosome between the two breaks gets deleted followed by the union of the broken ends of the chromosome. Such type of deletion is known as interstitial deletion.

For example, Prader-Willi syndrome and Angelman syndrome (interstitial deletion of long arm of chromosome 15).

Ring Chromosome (Fig. 4.2)

It is a type of deletion chromosome. If two breaks occur in a chromosome at its two ends, both the ends get lost, after which the two sticky broken ends unite to form a ring.

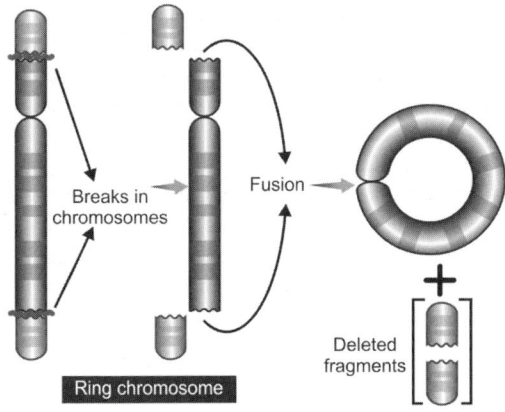

Fig. 4.2: Ring chromosome

Duplication

It is a process of addition of a portion of chromosome from another homologous chromosome with duplication of genes. Duplication may arise by unequal crossing over during meiosis or from rearrangement between two chromatids during mitosis.

Duplications are more common and less harmful to the individual than are deletions.

Translocation

Exchange of segments between nonhomologous chromosome is known as translocation. It requires breaks of both nonhomologous chromosomes, followed by repair leading to an abnormal arrangement.

A translocation may not always produce abnormal phenotype, but it can lead to formation of unbalanced gametes and carries a high-risk of abnormal progeny.

Translocation may be of two types:
1. Reciprocal translocation (Fig. 4.3A)
2. Robertsonian translocation (Fig. 4.3B).

Reciprocal translocation: It occurs between two pairs of non-homologous chromosomes, it may be *heterozygous* when only one of the chromosomes in a pair is involved, or *homozygous* when both members of a chromosome pair have exchanged segments with one another.

Robertsonian translocation (Centric fusion): It is a type of translocation in which the breaks occur at the centromeres of the two chromosomes and whole chromosome arms are exchanged. It usually involves two acrocentric chromosomes. For example in D/G translocation, the short arm of a D group chromosome (13–15) fuses with the short arm of a G group chromosome 21. The fragment formed by the fusion of the short arms of the two chromosomes is lost.

The incidence of Robertsonian translocation in man is about 1/1000 births. About 2% of all cases of Down's syndrome have Robertsonian translocation 14/21, 21/22 or rarely 21/21. Phenotypically, Robertsonian translocation carriers may be normal, but there is an increased risk of production of unbalanced gamets and therefore abnormal offspring.

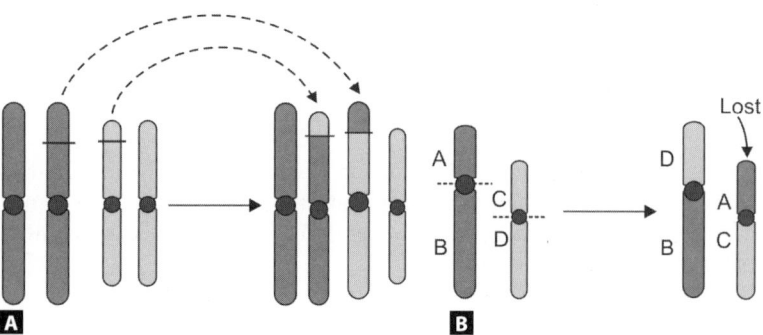

Figs 4.3A and B: (A) Reciprocal translocation; (B) Robertsonian translocation, showing fusion of two long arms and loss of the fused short arms of two chromosomes

Insertion

It is a nonreciprocal type of translocation, in which the chromosome segment drops out as a result of two breaks, become inserted at a single break in another chromosome. In insertion, there is merely a change in order of genetic material and hence no effect is seen on the phenotype of the individual.

Isochromosomes (Fig. 4.4)

The centromere of a chromosome, due to abnormal anaphase, splits transversely instead of longitudinal splitting. This results in formation of two chromosomes of unequal length, each presenting metacentric chromosomes with duplication of genes. The resulting chromosomes derived from transverse splitting of the centromere are known as isochromosomes.

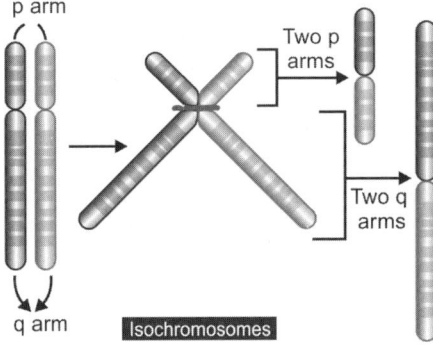

Fig. 4.4: Isochromosome

The isochromosome of long arm of X chromosome is the most commonly seen isochromosome, observed in some individuals with Turner's syndrome. Isochromosome 17q is seen in some patients with leukemia.

Inversion

A part of a chromosome is detached and later unites with the same chromosome in inverted position. The genes are not lost but placed in altered loci.

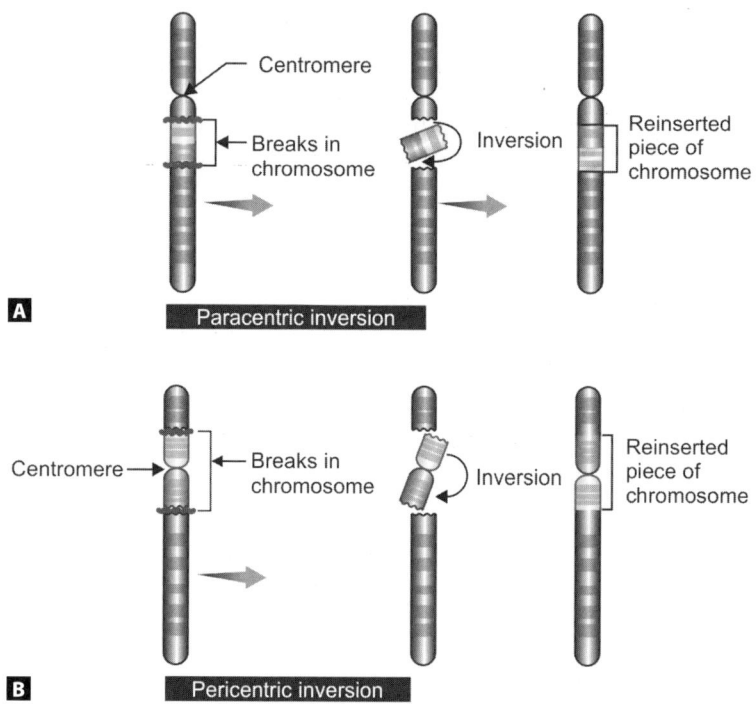

Figs 4.5A and B: Inversion

It is of two types:
1. Paracentric inversion (Fig. 4.5A)
2. Pericentric inversion (Fig. 4.5B).

Paracentric inversion: If the inverted segment does not include centromere.

Pericentric inversion: If the inverted segment includes centromere region.

■ NUMERICAL ABNORMALITIES

These may involve autosomes or sex chromosomes. The number of chromosomes per cell is fixed for a given species. In human beings, it is 46. This is called Diploid Number (2n). However, in gametes the number of chromosomes is only half the diploid number, i.e. 23. This is called Haploid Number (n). The basic cause responsible for aberrations in chromosomal number in almost all cases is nondisjunction during meiosis or mitosis.

Variations in the Chromosome Number

Polyploidy

It is used to denote presence of multiples of haploid number of chromosomes other than diploid number. Triploidy (3n) 69 and tetraploidy (4n) 92 are the two most commonly seen forms of polyploidy. Three sex types have been observed in triploidy, this being 69, XXX, 69, XXY and 69, XYY. Triploidy can result in abortions or in some cases live births that die at or shortly after birth. Tetraploids are usually 92, XXXX or 92, XXYY. Tetraploidy is very rare.

Aneuploidy

It is a condition where the chromosome number is altered by one or more, but not by multiples of haploid.
Aneuploidy can be of two types:
- *Hyperploidy*: It is the condition in which there is addition of one or more chromosomes to the diploid number. It may be called:
  - Trisomy (2n + 1) when one chromosome is added to the diploid number. Such a cell would have 47 chromosomes.
  - Tetrasomy (2n + 2) when two chromosomes of a homologous pair are added to the diploid number. Such a cell would have 48 chromosomes.
- *Hypoploidy*: It involves the loss of one or more chromosomes to the diploid number. It may be called:
  - Monosomy (2n – 1) when there is loss of one chromosome from the diploid set. A cell exhibiting this condition would contain 45 chromosomes.
  - Nullisomy (2n – 2) when both the chromosomes of a homologous pair are lost from the diploid set. Such a cell would have 44 chromosomes.

Symbols used in the cytogenetics:
- p—short arm of chromosome
- q—long arm of chromosome
- del—deletion
- t—translocation
- inv—inversion
- i—isochromosome
- r—ring chromosome
- s—satellite (for sat chromosome)

- mat—maternal origin
- pat—paternal origin
- slant line—indicates mosaicism (e.g. XY/XX , XO/XX)

+ or – sign when placed before an appropriate symbol, it means addition or missing of the whole chromosome. For example, trisomy 21 Down's syndrome is represented as 47, XY,+21.

When + or – signs are placed after a symbol, these indicate increase or decrease of the length of chromosome. For example, cri du chat syndrome with deletion of short arm of chromosome 5 is represented as 46, XY, 5p–.

In Philadelphia or Ph' chromosome, long arm of chromosome 22 is deleted and the deleted arm is then translocated to the long arm of chromosome 9. Symbolic representation of Ph' chromosome is 46, XY, t (22q–; 9q+).

Disorders Affecting Autosomes

- Cri du chat syndrome— 5p–
- Down's syndrome—Trisomy 21
- Patau's syndrome—Trisomy 13
- Edward's syndrome—Trisomy 18
- Philadelphia chromosome—t (22q–; 9q+).

Cri Du Chat Syndrome (Fig. 4.6)

Incidence: 1 in 50,000

General features
- Infant's cry is like mewing of a cat
- Growth retardation.

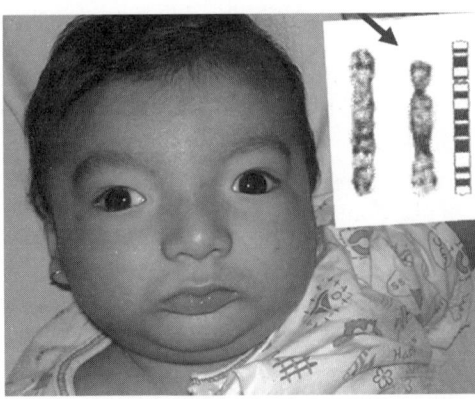

Fig. 4.6: Cri du chat syndrome

Mental status: Mental retardation
Head and neck:
- Microcephaly
- Moon like face
- Eyes widely spaced
- Micrognathia.

Cardiovascular system: Defects like ventricular septal defect may be seen.

Dermatoglyphics: Increased whorls and high ridge count.

Down's Syndrome (Mongolism) (Figs 4.7A to C)

Down's syndrome is the first chromosomal disorder to have been clinically defined and is the most commonly recognized genetic cause of mental retardation.

It is a chromosomal abnormality involving trisomy of autosomes.

History: It was first described by Langdon *Down* in 1866.

Incidence: It occurs once in every 800 to 1000 live births.

Predisposing factors: Its frequency increases with the increase of maternal age. It is probably due to increased risk of nondisjunction occurring in the aging ovary. Recent studies suggest that paternal age may also be related to the incidence of Down's syndrome.

Cause
- Trisomy 21 (95% of cases): Due to failure of separation of 21st pair of chromosomes during meiosis. This results in an extra chromosome on 21st pair.
- Translocation D/G: t(14q; 21q) Long arm of chromosome 21 is translocated to long arm of chromosome 14.
- Translocation 22/21
- Translocation 21/21.

Number of chromosomes in each cell: 47

Most prominent feature: Mental subnormality.

General Features
- Mental retardation (patients are in the IQ range of 25–50)
- Hypotonia or Lax limbs
- Short stature.

Head and Neck
- Small round head (Brachy-Cephaly)
- Oblique palpebral fissures

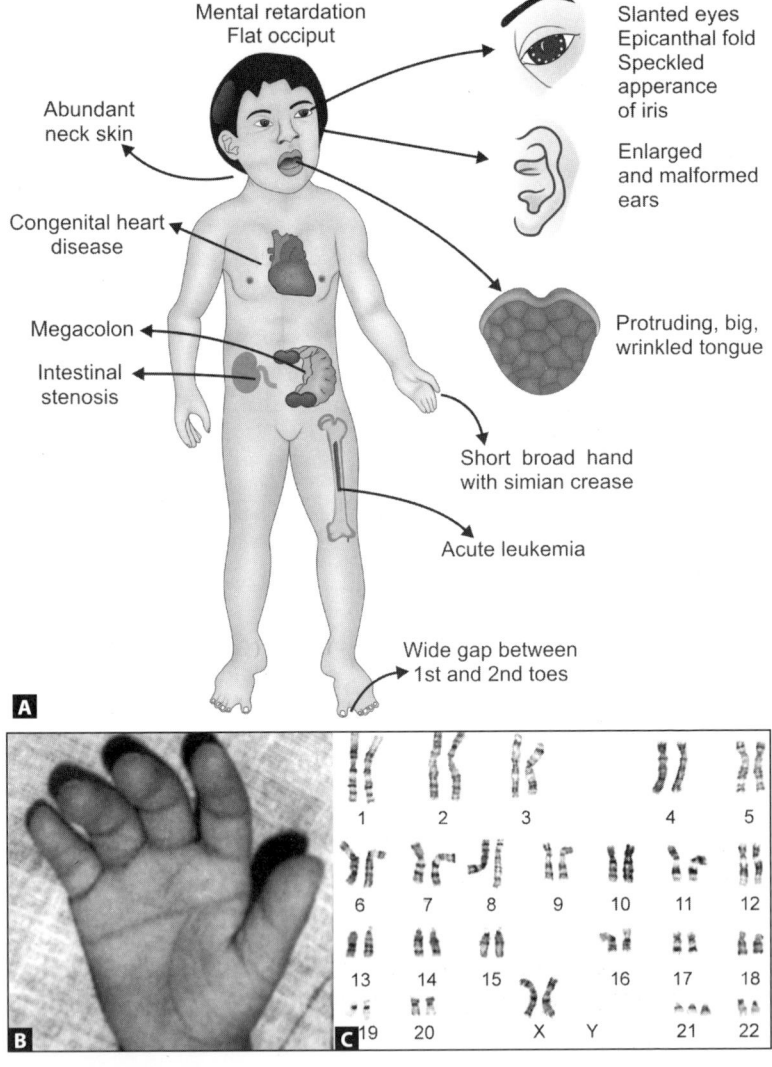

Figs 4.7A to C: Down's syndrome: (A) Major clinical features; (B) Palmar simian crease; (C) The karyotype

- Epicanthic folds
- Spots of depigmentation in iris
- Malformed ears

- Flat nasal bridge
- Furrowed lower lip
- Constantly open mouth with protruded and furrowed tongue
- Small teeth and delayed dentition
- High arched palate
- Broad and short neck
- *Hand*: Short, broad hand and short fingers
- *Feet*: Prominent gap between 1st and 2nd toes.

Cardiovascular system: Congenital cardiac defects, mainly in the form of ventricular septal defects and those involving atrioventricular canal.

Abdomen and Pelvis
- Umbilical hernia
- Duodenal atresia
- Cryptorchidism.

Dermatoglyphics
- Single transverse palmar crease (Simian crease)
- An atd angle greater than 57°.

Radiological Findings
- Hypoplasia of midphalanx of fifth finger
- Pelvic X-ray—iliac index is less than 60 degrees.

Patau's Syndrome (Fig. 4.8)

History: It was first identified by Patau and his collegues
Incidence: 1 in 10,000 births
Cause: Trisomy 13, (47, XY, +13).

Features
- Gross brain malformation
- Microphthalmia (Small abnormally formed eyes)
- Hair-lip or cleft palate
- Polydactyly
- Heart defects
- Renal abnormalities.

Survival rate
- About 95% of trisomy 13 conceptions are spontaneously aborted during pregnancy
- About 95% of live born infants die during the first year of the life.

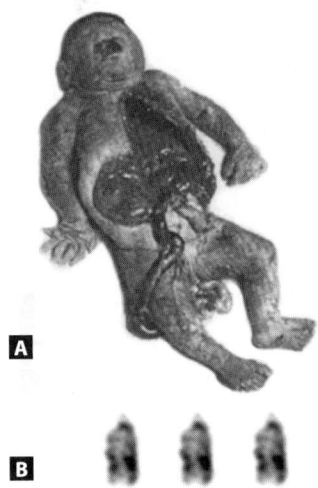

Figs 4.8A and B: Patau's syndrome (Trisomy 13): (A) Microcephaly, microphthalmos, cleft lip and palate and polydactyly; (B) Partial karyotype showing trisomy 13

Edward's Syndrome (Figs 4.9A to C)

Incidence: Second most common autosomal trisomy with a prevalence of about 1 per 6000 live births.

Cause: Trisomy 18 (47, XY, +18), an extra chromosome transmitted by the mother.

Features
- Elongated head
- Broad and flat nose
- Small, Low-set ears
- Micrognathia—small mouth which is hard to open
- Short sternum and short helluces (first toe)
- Overriding and contracture of fingers
- Hypermobility of shoulders
- Rocker bottom feet
- Congenital heart disease particularly ventricular septal defect in 90% of children.

Survival rate: It is the most common chromosomal abnormality among stillborns with congenital malformations.

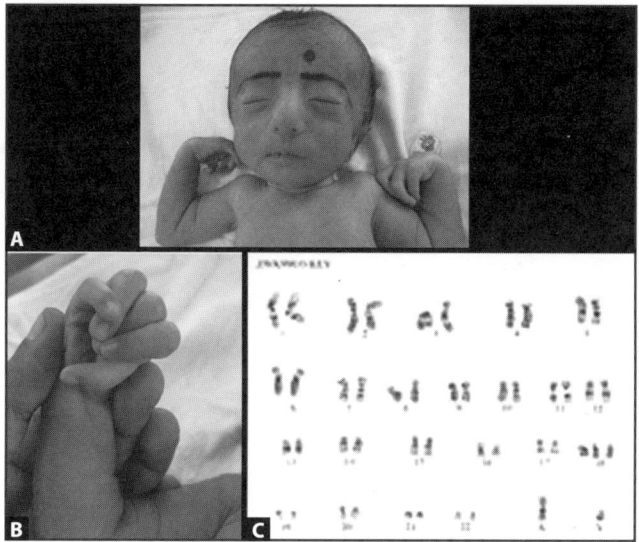

Figs 4.9A to C: Edward's syndrome (Trisomy 18)

Less than 5% of trisomy 18 conceptions survive to term and die within the first several weeks of life.

Philadelphia Chromosome

This abnormality is found in some cases of chronic myeloid leukemia (Fig. 4.10).

History: It was first read in Philadelphia conference in 1960.

Cause: It is due to reciprocal translocation of long arm of chromosome 22 on to the long arm of chromosome 9. A small distal portion of long arm of chromosome 9 in turn is translocated to chromosome 22. The net effect is a smaller chromosome 22, which is known as Philadelphia chromosome. It is represented as t (22q–; 9q+).

Cytogenetic findings: About 85% of diagnosed chronic myeloid leukemia patients are Ph' positive (Philadelphia chromosome positive). It is found in blood and bone marrow cells, but other tissues usually shows normal chromosome complement.

The presence of Ph' chromosome is the result rather than the cause of the disease.

48 Basic Human Genetics

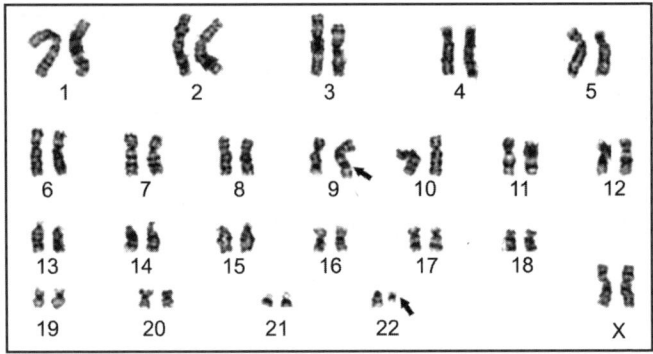

Fig. 4.10: Karyotype of patient with chronic myeloid leukemia

■ DISORDERS AFFECTING SEX CHROMOSOMES

- Turner's syndrome
- Klinefelter's syndrome
- Triple X syndrome
- XYY syndrome
- 46, XY female (Testicular feminization)
- 46, XX male
- Fragile X syndrome
- Intersex.

Turner's Syndrome (Fig. 4.11)

It is a most common sex chromosomal abnormality. It is monosomy of X chromosome.

History: It was described by Henry Turner in 1938.

Incidence: It occurs 1 in 2000 to 1 in 3000 live born girls.

Chromosomal complement: Most commonly 45, XO with complete absence of one X chromosome.
Sometimes XX/XO mosaicism is observed. Barr body is negative.

Phenotype: Female

Most prominent feature: Amenorrhea

General Features
- Short stature (most constant feature)
- Amenorrhea (primary)

Chromosomal Abnormalities

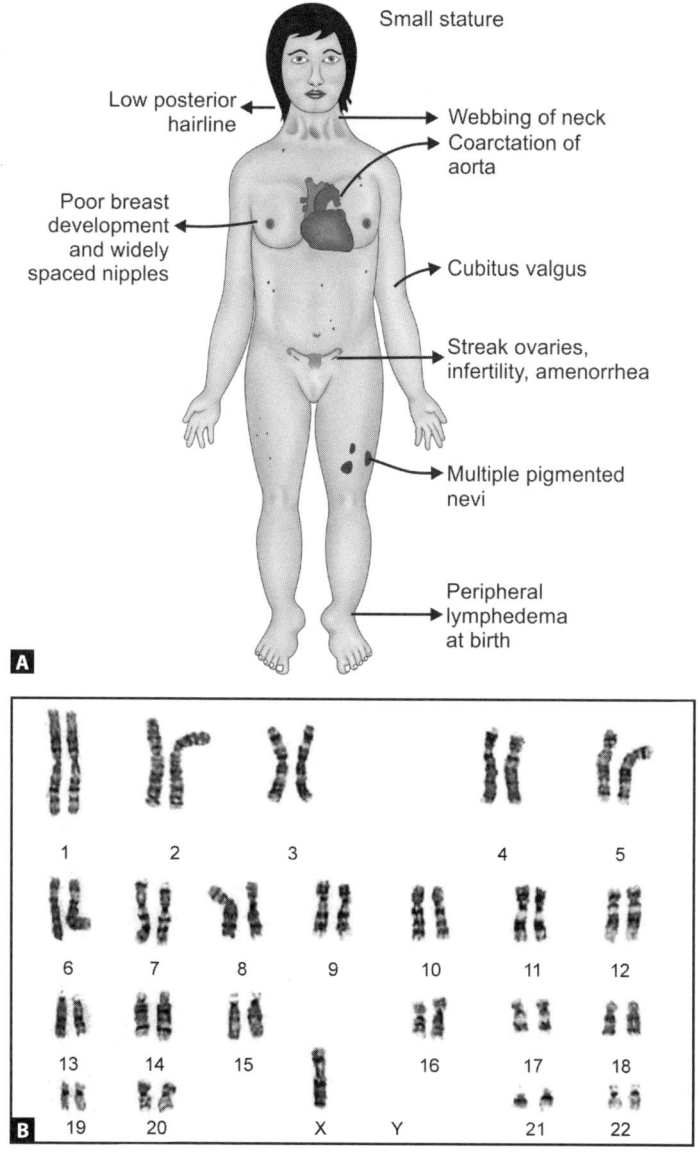

Figs 4.11A and B: Turner's syndrome: (A) Features; (B) Karyotype

- Sterility
- Sexual infantilism
- Scanty pubic or axillary hair.

Head and Neck
- Epicanthal folds
- Defective vision
- Small mandible
- Narrow maxilla
- High palate
- Short and webbed neck
- Low posterior hair line.

Thorax
- Broad chest
- Underdeveloped breasts.

Cardiovascular System
- Coarctation of aorta
- Aortic stenosis
- Pulmonary stenosis.

Abdomen and Pelvis
- Urinary tract anomalies (especially involving kidneys)
- Absence of ovaries (Gonadal dysgenesis). At laparotomy, ovary is found to consist of undifferentiated stroma with no sex cells. A pale strip of fibrous tissue is attached to the back of the broad ligament. This is sometimes called as "Streak" gonad.
- Vagina and uterus are under or maldeveloped.

Endocrinological defects: Like diabetes mellitus and Cushing's syndrome.

Limbs
- Peripheral lymphedema
- Cubitus valgus
- Dystrophic nails.

Dermatoglyphics: Raised ridge count.

X-ray findings
- Hypoplasia of lateral ends of clavicles
- Positive metacarpal sign (short fourth and fifth metacarpals).

Investigations required
- *Chromosomal analysis*: It shows chromosomal complement to be XO. Barr body is absent.

- Plasma levels of gonadotropins, especially of follicle-stimulating hormone (FSH) are generally elevated.
- Radiological studies may show renal or cardiovascular abnormalities.

Treatment

Replacement therapy is given with estrogens to initiate and maintain sexual maturation. This therapy may be postponed until the age of puberty, in order to prevent early closure of epiphyses.

Klinefelter's Syndrome (Fig. 4.12)

It is a chromosomal disorder involving trisomy of sex chromosomes.

History: This condition was first recognized by Harry Klinefelter and Co-workers in 1942.

Incidence: 1 in 500 to 1,000 male births.

Cause: This chromosomal aberration may be the result of meiotic non-disjunction of an X chromosome during gametogenesis or from mitotic nondisjunction in zygote.

Chromosome Complement

Majority of cases have 47, XXY chromosomal complement. In about 15% of cases mosaicism is found, i.e. 46 XY/47 XXY. The extra X chromosome is derived maternally in about 50% of Klinefelter cases and the incidence of the syndrome increases with advanced maternal age. Variants of Klinefelter's syndrome such as 48, XXXY or 48, XXYY or 49, XXXXY shows additional X chromosome with severe mental retardation. Barr body is positive in phenotypic male.

Phenotype: Male

Most prominent feature: Gynecomastia and underdeveloped male sexual characters.

Features

It is difficult to diagnose Klinefelter's syndrome before puberty because the features are not so prominent in childhood.

The important features are:

General
- Tall with disproportionately long lower limbs
- Scanty growth of hair
- Gynecomastia.

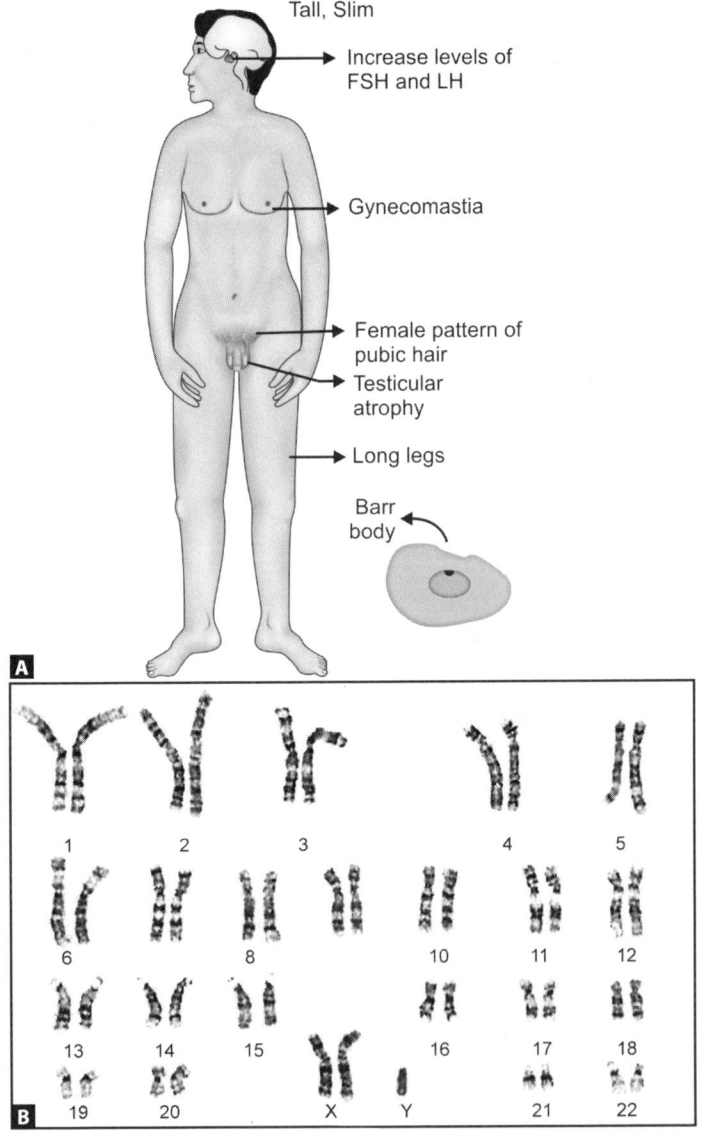

Figs 4.12A and B: Klinefelter's syndrome: (A) Features; (B) Karyotype
Abbreviations: FSH, follicle-stimulating hormone; LH, luteinizing hormone

Note the male genitalia, gynecomastia and tendency towards female distribution of pubic hair

Mental Status
- Intellectual development is fair to good.

Genital System
- Small testes
- Azospermia
- Infertility
- Underdeveloped male secondary sex characters.

Associated Features

There is an increased incidence of:
- Pulmonary diseases
- Varicose veins
- Cancer of breast.

Dermatoglyphics
- Excess of Arches with a low ridge count

Investigations

- Chromosome analysis reveals trisomy of sex chromosomes. Buccal smear should be examined for Barr bodies which indicate the number of X chromosomes
- Plasma testosterone levels are low
- Gonadotropin levels are usually elevated by the time of puberty
- Testicular biopsy: Before puberty, only deficiency or absence of germinal cells is seen. After puberty, it shows hyaline degeneration of seminiferous tubules.

■ TREATMENT

Replacement therapy with long-acting testosterone preparation should begin at the age of 11–12 years.

Triple X Syndrome

- *Chromosomal complement*: 47, XXX
- *Incidence*: 1 in 1000 females
- *Phenotype*: female with presence of 2 Barr bodies.
- *Features*: Usually, no marked abnormalities are seen. Sometimes, there may be mild mental retardation, menstrual irregularities or sterility are observed. Females with triple X syndrome may have normal children.

XYY Syndrome

- *Chromosome complement*: 47, XYY
- *Incidence*: 1 in 1000 males
- *Phenotype*: Male.

Features
- Taller than average (above 6 ft)
- 10–15 point reduction in average IQ
- Unsound mind, aggressive and antisocial behavior
- Learning disabilities.

46, XX Males

The affected individual is phenotypically male with gynecomastia, infertility and testicular atrophy. It may be due to one of the X chromosome carrying Y bearing genes from paternal side during crossing over in gametogenesis.

46, XY Females (Testicular Feminization Syndrome)

The phenotype of the individual is female, but the karyotype is 46,XY. It occurs due to androgen insensitivity. The affected female possesses fully formed abdominal testes. The molecular defect is in SRY gene located on short arm of Y chromosome, which regulates the development of testes.

Fragile X Syndrome

It is caused by mutation involving one X chromosome of the female and so it is transmitted to her male child. The affected boys are mentally retarded and have a tendency to possess large ears, protruding chin, large hands and feet and large testes.

■ INTERSEX

An intersex is defined as an individual with ambiguous internal and external genitalia. Intersex conditions are not always due to aberrations of sex chromosomes. They can also be due to single mutant genes. They are clinically of two varieties, true hermaphroditism and pseudo-hermaphroditism. This classification is based on nature of gonads.

True Hermaphrodite

Person having presence of both male and female gonadal tissue, as mixed gonad (ovo-testis) or alternatively an ovary on one side and testicle on other side.

Pseudohermaphrodite

Person showing normal gonadal sex (testis in male and ovary in female) and normal sex chromosomal constitution (male with XY and female with XX) but varying degree of abnormal external and/or internal genitalia, towards the opposite sex.

Female pseudohermaphrodite: Person having normal ovary and internal genitalia, but external genitalia and secondary sexual characters resembling those of male.

Male pseudohermaphrodite: Person having functional or nonfunctional testes with internal and/or external genitalia resembling those of female.

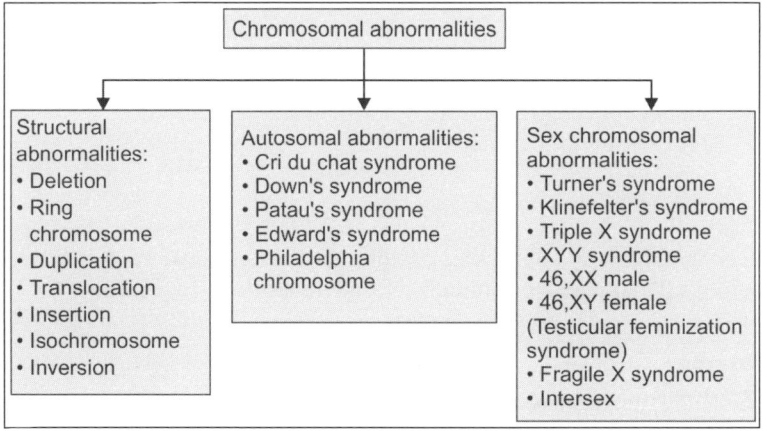

5

Genes

■ DEFINITION

Genes are the structural units of inheritance stored in chromosomes.

History

The term 'gene' was introduced by Johannsen in 1909.

Properties

Genes possess the following properties:
- Capability to determine traits, e.g. skin color, intelligence, temperament, blood group, etc.
- Ability to undergo identical reproduction (Replication)
- Ability to undergo mutation.

Location

Every gene lies in a particular position on the chromosome to which it belongs. This specific position is referred to as the Locus of Gene.

Number

Nucleus of each cell of human body contains about 20,000 genes, half of which being contributed by each parent.

■ CLASSIFICATION

- *Structural gene (Cistron)*: It controls the structure of a polypeptide comprising the protein.
- *Super gene (Polycistron)*: It is a group of genes which encode a number of functionally related proteins.

- *Operator gene*: It exercises control over a set of contiguous structural genes and collectively constitutes an operon.
- *Regulator gene*: It is responsible for controlling the particular activity of a specific operon.
- *Temporal gene*: It is concerned with regulation of protein synthesis in relation to time.
- *Architectural gene*: It guides the exact localization of enzymes and proteins in various cellular organelles.
- *Pleiotropic gene*: Such a gene influences more than one phenotypic trait, e.g. gene responsible for causing the disease Phenyl Ketonuria, also results in other abnormalities like mental retardation and short stature. This phenomenon of multiple expression by the genes is referred to as Pleiotropism.
- *Sex-linked genes*: These are genes linked with sex chromosomes. Most sex-linked genes are linked with X-chromosomes. Few Y-linked genes (Holandric genes) have also been described, e.g. one associated with hairy pinna.

Functions

- Genes maintain the specificity of an individual
- Genes play vital role in the transmission of characters from parents to offsprings
- Genes are very important for synthesis of various proteins and enzymes of the cell.

■ EFFECTS OF GENE

Profound variability is seen in relation to gene effects. The phenotypic effect produced by a gene is influenced by:
- Its interaction with other genes
- Environmental factors.

The following terms are used to describe the gene effects:

Penetrance: It denotes the frequency with which a particular gene produces its effect in an individual.

Expressivity: The term indicates the degree of effect produced by a specific gene in different individuals.

■ CHEMICAL BASIS OF GENE (DNA)

DNA is the chemical basis of genes and is, indeed, the hereditary material. Genes act through controlling the protein constitution of cells.

Amount: In man, each somatic cell nucleus contains about 5.6×10^{-12} g of DNA.

Structure: DNA is arranged in the form of fine fibers. Two parallel strands are entwined spirally to constitute a Double Helix (Fig. 5.1). Each molecule can be simulated to a spiral ladder in which sides are formed of sugar and phosphate, whereas each stair is made by a pair of nitrogenous bases. Each strand of DNA molecule is made up of nucleotides, arranged in the form of chains. Basically, a nucleotide comprises of (Fig. 5.2):

- A sugar—Deoxyribose
- A molecule of phosphate
- A nitrogenous base, which can be one of the following: Adenine — Thymine — Guanine — Cytosine.

The two strands are held together by hydrogen bonds between the bases of respective strands. This linkage takes place in a manner that adenine always links with thymine, whereas guanine is always

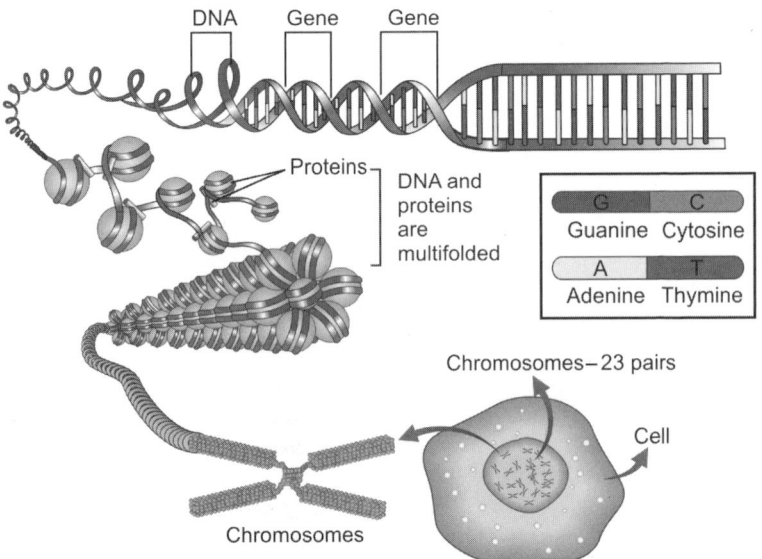

Fig. 5.1: Structure of chromosome, DNA and gene

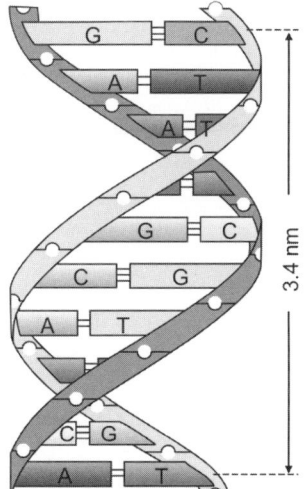

Fig. 5.2: Double helical structure of DNA

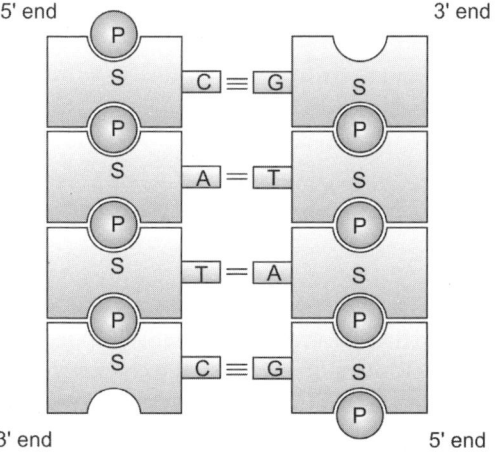

Fig. 5.3 : Basic structure of nucleic acid

connected with cytosine (Fig. 5.3). This specific arrangement of bases is referred to as complementary base pairing.

Nucleic Acids

There are two types of nucleic acids:
1. DNA (Deoxy-Ribose Nucleic Acid)
2. RNA (Ribose Nucleic Acid).

The two types of nucleic acids are compared in the Table 5.1.

Table 5.1: Comparison of DNA and RNA

Features	DNA	RNA
1. Site	Forms the chemical basis of genes stored in chromosomes	Chiefly concentrated in cytoplasm and also in nucleolus
2. Appearance	Double stranded structure	Single stranded structure
3. Form	Single	Exists in 3 forms: 1. Messenger RNA (mRNA) 2. Transfer RNA (tRNA) 3. Ribosomal RNA (rRNA)
4. Sugar molecule	Deoxyribose	Ribose
5. Bases	Adenine, thymine, guanine and cytosine	Adenine, uracil, guanine and cytosine
6. Reaction with alkali	Cannot be hydrolyzed by alkali	Can be hydrolyzed by alkali
7. Function	It is the hereditary material	It takes part in protein synthesis

■ MUTATIONS

Definition

This term is used to denote a sudden heritable physicochemical change in DNA (gene) which alters the effect on the character influenced by it.

History

The first mutation was recognized by TH Morgan in Drosophila, in 1910.

Classification

Mutations can be classified in different ways.
- On the basis of genes involved:
    - *Point mutations*: These mutations involve single gene.

- *Gross or chromosomal mutation*: These mutations involve a chromosome or a set of chromosomes, e.g. polyploidy.
- On the basis of origin:
  - *Natural*: These mutations arise spontaneously from natural causes.
  - *Induced*: These mutations arise as a result of man-made agents, e.g. radiations, chemicals.
- On the basis of type of change:
  - *Structural:* These mutations involve changes in the nucleotide content of gene, e.g.
    - *Deletion mutation*: A portion of gene is lost in this type of mutation. Small deletions may remove one or a few base pairs within a gene, while larger deletion can remove an entire gene. The deleted DNA may alter the function of the resulting proteins (Fig. 5.4).
    - *Insertion mutation*: This mutation involves addition of one more extra nucleotide to a gene. As a result protein made by the gene may not function properly (Fig. 5.5).
    - *Duplication mutation*: A duplication consists of a piece of DNA which is abnormally copied one or more times.
    - *Frameshift mutation*: This mutation occurs when the addition or loss of DNA bases change a gene's reading frame. A reading frame consists of groups of three bases that each code for one amino acid. A frameshift mutation shifts these bases and changes the code for amino acids (Fig. 5.6).

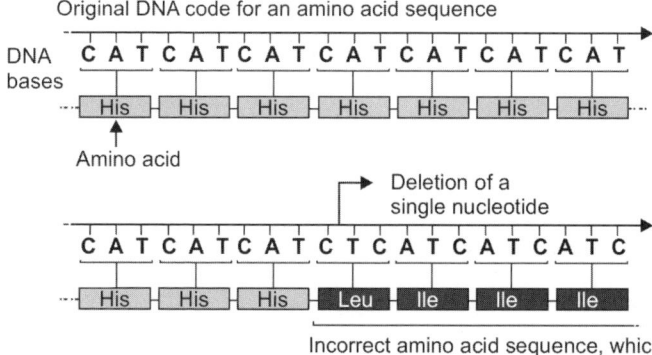

Fig. 5.4: Deletion mutation

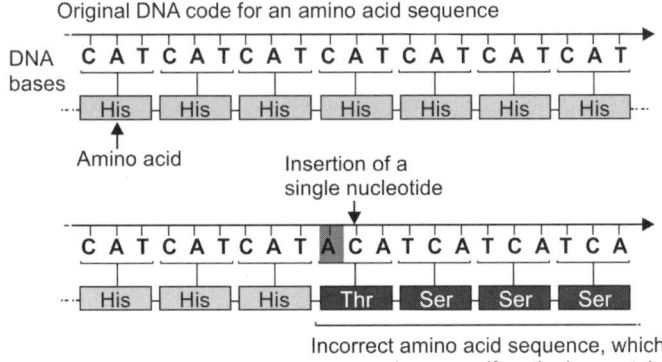

Fig. 5.5: Insertion mutation

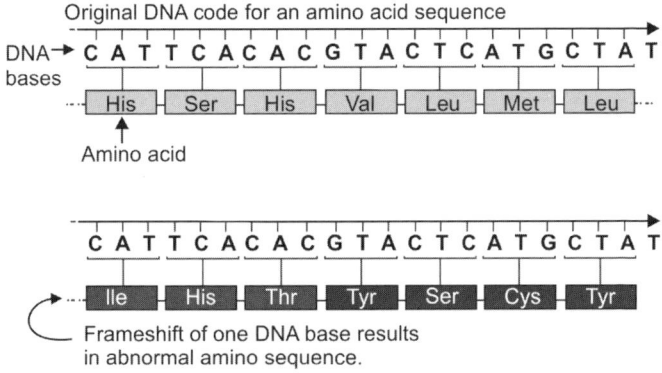

Fig. 5.6: Frameshift mutation

- *Missense mutation*: It involves a change in one DNA base pair that results in the substitution of one amino acid for another in the protein made by a gene (Fig. 5.7).
- *Nonsense mutation*: It involves substitution of one amino acid for another, so that the alter DNA sequence signals the cell to stop forming the protein. Such protein functions improperly or not at all (Fig. 5.8).
- *Rearrangement*: These types of mutations involve change of location of a gene within the genome.

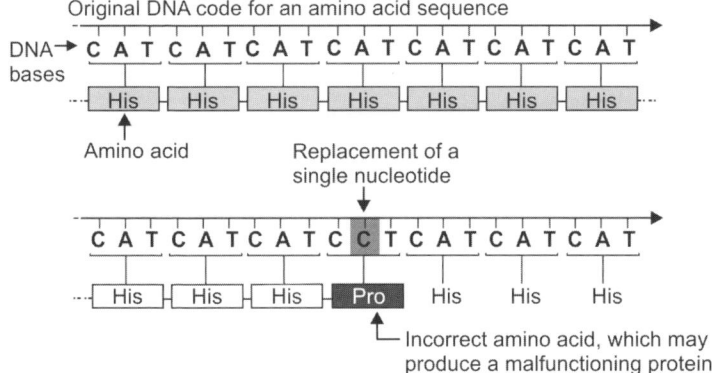

Fig. 5.7: Missense mutation

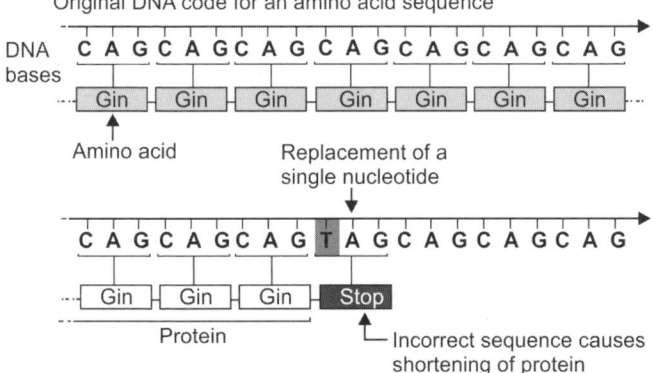

Fig. 5.8: Nonsense mutation

- On the basis of cell type:
  - *Somatic cell mutation:* These mutations involve the somatic cells of body. Such mutations are not transmitted to the offspring. However, localized changes may occur and even malignant growths may be formed.
  - *Germ cell mutation*: These mutations occur in the reproductive cells. Such mutations can be perpetuated in the subsequent generations. Effect of mutation seen in subsequent generations

will depend on the type of character involved. If it is dominant, the effect is seen in the next generation. However, if the character is recessive, several generations may escape before it is observed.
- On the basis of direction
  - *Forward*: This type of mutation creates a change from wild type (original) to an abnormal phenotype.
  - *Reverse*: Such mutations produce a change from mutant type to wild type.

Mutagens: These are factors capable of causing mutation.

Some of these are:
- X-rays
- Ultraviolet rays
- Other radiations, e.g.
  - Alpha rays
  - Beta rays
  - Cosmic rays.
- Temperature.
- Chemicals, e.g.
  - Mustard gas
  - Phenol
  - Caffeine
  - Formaldehyde
  - Nitrous acid
  - Acridine.

Significance

Although most mutations being harmful to the organism concerned, yet are of significance with respect to the following:
- *Evolution of species*: Accumulation of advantageous mutations are useful in bringing about evolution of species.
- *Understanding the basic principles of inheritance*: Mutations are of tremendous help in understanding the basic principles of inheritance.
- *Study of cell metabolism*: Mutations are utilized for understanding mechanisms of cell metabolism, e.g. biochemical pathways
- *Improvement of synthesis of antibiotics from microorganisms*: Various mutant strains of microorganisms with better yields of antibiotics are being widely used in different industries. One such application of induced mutations has been in the improvement of yield of Penicillin by the mold *Penicilium*.

6
Inheritance

The entire human body develops from one cell, the fertilized ovum which carries all the necessary information for the formation of numerous tissues and organs of the body. This tremendous volume of information is stored in the chromosomes of each cell. Each chromosome carries on itself large number of genes.

Genes are the structural units of inheritance stored in chromosomes, and are responsible for transmission of characters from parents to offspring. Genes, like chromosomes, are arranged in pairs. During gametogenesis, two chromosomes of a pair along with their genes separate and pass on to different gametes in a purely random manner. Therefore, different gametes have different combinations of chromosomes and genes. When these gametes take part in fertilization, new chromosome and gene pairs are formed.

■ INHERITANCE

It is a term used to denote transmission of characters from parents to offspring.

We have seen in the previous chapter that every gene has a specific position on the chromosome to which it belongs. This position is termed as the locus of gene. However, genes at a locus may assume two or more forms, each possessing a different effect on the character concerned.

For example, let us study the genes that control the growth of nails. The normal gene controls the proper growth of nails, whereas the abnormal one exerts an inhibitory effect. In an individual, there can be three possibilities:
1. Both normal genes (Homozygous normal)
2. Both abnormal genes (Homozygous abnormal)
3. One gene normal and other abnormal (Heterozygous).

In a homozygous normal individual there would be normal growth of nails. A homozygous abnormal individual will show anonychia.

In a heterozygous individual also, anonychia is seen. This is because the abnormal gene dominates the normal gene and is thus expressed. This kind of inheritance is known as dominant inheritance.

In certain other conditions, e.g. phenylketonuria, the disease does not manifest in the heterozygous individual. It is seen only in a homozygous individual carrying two abnormal genes. Here, the abnormal gene remains suppressed in a heterozygous individual. This kind of inheritance is known as recessive inheritance.

Genes carried on chromosomes are responsible for the development of inherited characters or traits. When a single gene pair determines a trait, the trait is known as single gene trait and its mode of inheritance from one generation to next generation follows Mendel's law of unit inheritance and segregation. When many genes determine a trait, the trait is known as polygenic and it follows the pattern of polygenic inheritance.

■ MENDEL'S LAWS OF INHERITANCE

Gregor Mendel in 1865 worked independently on garden peas (pisum sativum) and laid the foundation stone of modern genetics. He was the pioneer in discovering and postulating the theories of modern genetics for which he is now recognized as *The Father of Modern Genetics* and his theories are being recognized as Mendelism.

He selected seven pairs of contrasting characters in the garden peas, such as height of plant, shape of pod, texture of seed, flower color and position, etc.

He confined his attention to a single character at a time as for example, height of the plant (tall and short). After completing his observation for one trait (height of the plant), he also studied other traits one by one. He also counted the numbers of each type of progeny in each cross.

Mendel crossed tall pea plant with short pea plant and observed that all the plants in the first generation were tall. He symbolized first generation plants as F1 generation. He explained that the shortness of the short parent was suppressed and that the tallness dominated. So, he designated tallness as dominant in contrast to shortness as recessive character in F1 generation plants (Fig. 6.1).

Subsequently, Mendel crossed the F1 generation plants amongst themselves by allowing self fertilization. He observed that second generation plant (F2) were of two types few were short and few were tall plants.

The ratio between tall and short plants in F2 generation was 3:1 (Fig. 6.2).

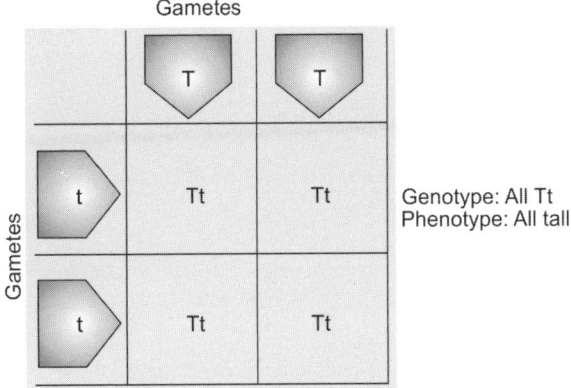

Fig. 6.1: Punnett square showing offspring resulting due to mating between the pure tall and pure short plants

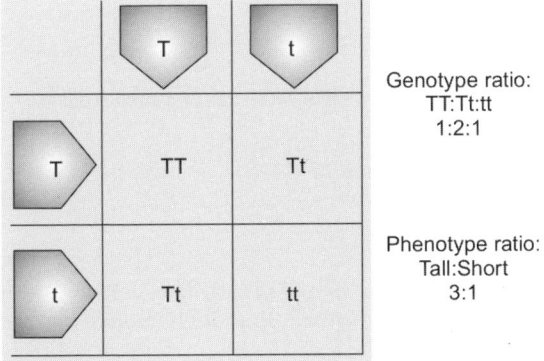

Fig. 6.2: Offspring resulting due to self-pollination of F1 generation

Mendel's Laws of Inheritance

Law of Unit Characters

This law states that in the zygote, two units of characters one from mother and one from father remain independent and maintain their individual identity. They do not blend or react to each other; if they do not express in first generation they can reappear without change in subsequent generation. In above stated example cross between tall and short plants

68 Basic Human Genetics

Mendel's experiment:
Tall character = T
Short character = t
First generation = F1
Second generation = F2

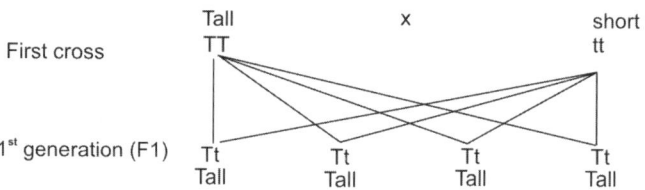

All plants in F1 generation are heterozygous tall

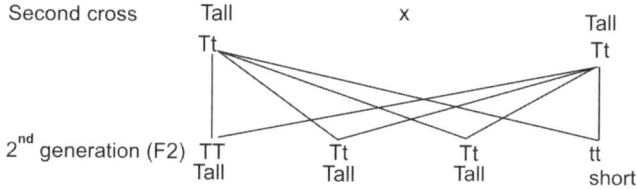

In F2 generation the ratio between tall and short plant is 3 : 1

TT = Homozygous tall (25%)
Tt = Heterozygous tall (50%)
tt = Homozygous short (25%)

led to F1 generation having all tall plants. The shortness reappeared in F2 generation. There was no blending of characters like tall + short = medium height plant.

Law of Dominance

The factors or genes exist in pairs. The character that is expressed in the first generation is called *dominant* and the character that remains suppressed in the first generation is called *recessive*.

Law of Segregation

The two alternative factors for each trait get separated during gametogenesis and each factor has an equal chance of getting transferred to the offspring.

Law of Independent Assortment

The inheritance of one factor is unaffected by the inheritance of the other. Each gene gets assorted independently of the other during its passage from one generation to the other. For example, a tall plant with yellow seed and a short plant with green seed are crossed, in F1 generation only the dominant characters such as tallness and yellow color of the seed will be expressed in the offsprings. When cross breeding is done between offsprings of F1 generation, four different varieties—tall yellow, short yellow, tall green, short green are observed in the ratio of 9:3:3:1.

Few Terms (Fig. 6.3)

- *Alleles*: Alternative forms of the same gene are termed alleles. Each individual carries two sets of genes, one from each parent.
- *Homozygous*: An individual is said to be homozygous when the two members of a pair of genes are alike.
- *Heterozygous*: An individual is said to be heterozygous when the two members of a pair of genes are different.
- *Genotype*: It is the term used for genetic constitution of an individual.
- *Phenotype*: This term is used to indicate the physical manifestations of characters like form, size, color, etc.

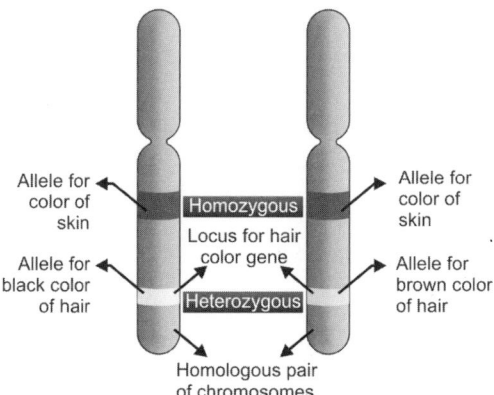

Fig. 6.3: Chromosome to show allele

Pedigree Analysis

To study a particular trait in a family, the data are collected from a family over a number of generations and represented in a chart using international symbols. Such charts are known as pedigree charts.

The first person of a family in whom the trait appears first is known as index case or proband (such person is also known as propositus if male, proposita if female).

Symbols used for pedigree analysis are shown in Figure 6.4:

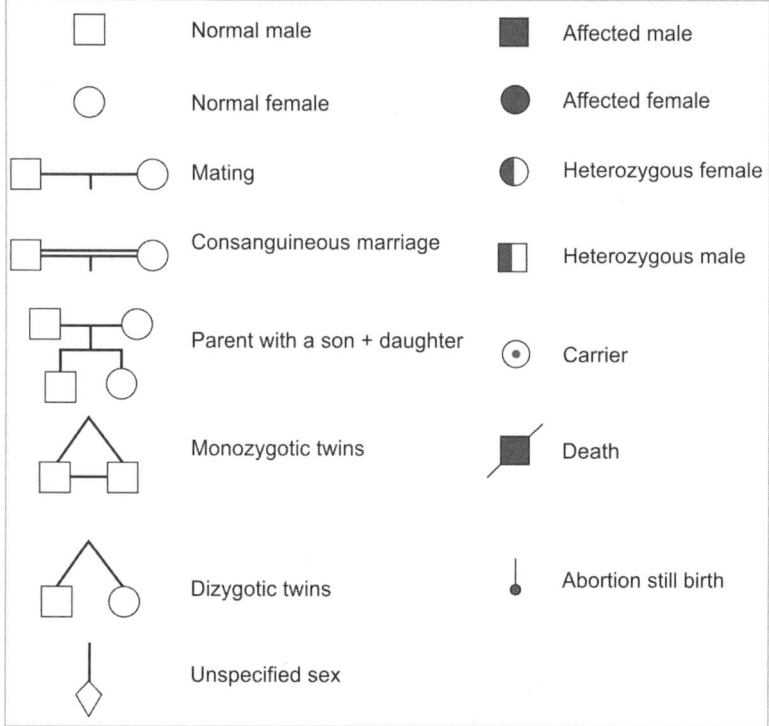

Fig. 6.4: Symbols used in pedigree chart

■ SINGLE GENE INHERITANCE

The gene responsible for a disorder or a trait if located on an autosome, it is said to follow autosomal inheritance and if located on a sex chromosomes is said to follow a sex linked inheritance.

These single gene inheritance patterns are further classified into:
- Autosomal dominant inheritance
- Autosomal recessive inheritance

- Sex-linked inheritance
  - X-linked dominant inheritance
  - X-linked recessive inheritance
  - Y- linked inheritance.

Autosomal Dominant inheritance (Fig. 6.5)

If a trait manifest itself in a heterozygous state, only one copy of mutant gene is needed for manifestation of disease. This means affected person carries a single copy of the affected gene, and the other copy is normal. Incidence—1 in every 200 persons.

Characteristics

- Abnormal gene dominates the normal gene, and is thus expressed in heterozygous state.
- Affected person is either homozygous or heterozygous.
- Unaffected parent do not transmit the trait to their offsprings.
- Normal children of an affected person do not transmit the disease.
- Disease can be inherited from one parent also.
- The trait appears in every generation without skipping, it has a vertical transmission and there is always at least one affected parent.
- It is not related to consanguineous marriage.
- Males and females have equal chances of getting and transmitting the disease.

Examples

- Achondroplasia
- Osteogenesis imperfecta

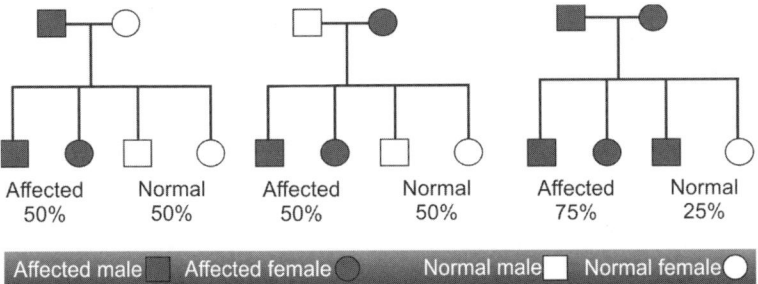

Fig. 6.5: Autosomal dominant inheritance

Basic Human Genetics

- Brachydactyly
- Anonychia.

Let us now see what happens when persons of different genetic constitutions marry. We shall use signs + and – to depict normal and abnormal genes respectively. The possibilities of various genetic combinations are depicted in Table 6.1.

Table 6.1: Autosomal dominant inheritance—genetic constitution of offsprings in various situations

Sr. No.	Genes of parents	Offsprings	Results
1.	Homozygous normal ++ Homozygous normal ++	++, ++, ++, ++	All children are homozygous normal
2.	Homozygous abnormal – – Homozygous abnormal – –	– –, – –, – –, – –	All children are homozygous abnormal
3.	Heterozygous + – Heterozygous + –	++, + –, + –, – –	Homozygous normal—25% Heterozygous abnormal—50% Homozygous abnormal—25%
4.	One parent heterozygous + – One parent homozygous abnormal – –	+ –, + –, – –, – –	Heterozygous abnormal—50% Homozygous abnormal—50% All children suffer from the disease
5.	One parent heterozygous + – One parent homozygous normal ++	++, ++, +–, +–	Heterozygous abnormal —50% Homozygous normal —50%

Autosomal Recessive Inheritance (Fig. 6.6)

The recessive trait is expressed only in homozygous state. Heterozygous are carriers of the disorder.

Incidence—2.5 per 1000 births.

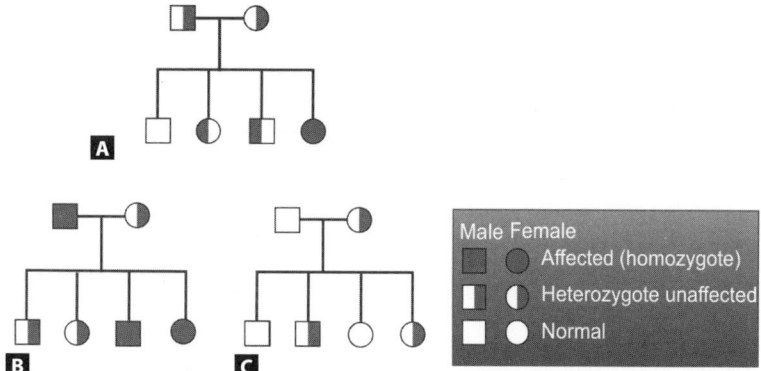

Figs 6.6A to C: Pedigree illustrating mechanism of autosomal recessive transmission. (A) both parents are unaffected heterozygotes; (B) One parent is sufferer and other is unaffected heterozygote; (C) One parent is normal and other is an unaffected heterozygote

Characteristics

- Abnormal gene remains suppressed and does not manifest in a heterozygous individual.
- Affected person is homozygous for the trait.
- Unaffected parents can transmit the trait to their offsprings, if they are carrier.
- Disease can appear only if both parents transmit it.
- The disease has horizontal pattern of transmission, siblings are affected.
- The abnormality is more commonly seen in consanguineous marriages.
- Males and females have equal chances of getting and transmitting the disease.

Examples

- Thalassemia
- Cystic fibrosis
- Sickle cell anemia
- Phenylketonuria.

The possibilities of various genetic combinations are depicted in Table 6.2.

Basic Human Genetics

Table 6.2: Autosomal recessive—genetic constitution of offsprings in various situations

Sr. No.	Genes of parents	Offsprings	Results
1.	Homozygous normal ++ Homozygous normal ++	++, ++, ++, ++	All children are homozygous normal
2.	Homozygous abnormal - - Homozygous abnormal - -	- -, - -, - -, - -	All children are homozygous abnormal
3.	Heterozygous + - Heterozygous + -	++, + -, + -, - -	Homozygous normal— 25% Heterozygous carrier—50% Homozygous abnormal—25%
4.	One parent heterozygous + - One parent homozygous abnormal - -	+ -, + -, --, - -	Heterozygous carrier—50% Homozygous abnormal—50%
5.	One parent heterozygous + - One parent homozygous normal ++	++, ++, +-, +-	Heterozygous carrier—50% Homozygous normal—50% None manifest the disease

To summarize autosomal dominant and autosomal recessive inheritance, see Table 6.3 of comparison.

Table 6.3: Comparison of autosomal dominant and autosomal recessive inheritance

Autosomal dominant inheritance	Autosomal recessive inheritance
1. Abnormal gene dominates the normal gene and is thus expressed in a heterozygous individual	Abnormal gene remains suppressed and does not manifest in a heterozygous individual
2. Affected person is homozygous or heterozygous	Affected person is homozygous for the trait
3. Unaffected parents do not transmit the trait to their off springs	Unaffected parents can transmit the trait to their offsprings (if both parents carry a recessive gene)

Contd...

Contd...

4.	Disease can be inherited from one parent also	Disease can appear only if both parents transmit it
5.	Members are affected in each generation. There is always at least one affected parent	Siblings may be affected but parents, usually are apparently normal
6.	Such diseases are known as hereditary diseases	Such diseases are known as familial diseases
7.	It is not related to consanguineous marriages	The abnormality is more commonly seen in consanguineous marriages
8.	If both parents are heterozygous, 75% of offsprings manifest the disease	If both parents are heterozygous, 25% of offsprings manifest the disease
9.	If one parent is homozygous normal and the other is heterozygous, 50% of the offsprings suffer from disease	If one parent is homozygous normal and the other is heterozygous, 50% of offsprings are carriers, but none manifests the disease
10.	If one parent is homozygous abnormal and the other is heterozygous, all the offsprings suffer from disease	If one parent is homozygous abnormal and the other is heterozygous, 50% of the offsprings suffer from disease

Sex-linked Inheritance

Sex-linked inheritance is a type of inheritance occurring as a result of mutant genes located on the X or Y chromosomes. The disorders, which occur due to mutant genes located on X chromosomes, are known as X-linked disorders. There is no such genes on Y chromosome, but certain trait like hairy pinna is the only known condition linked with Y chromosome. This is called holandric inheritance. In short, it means that sex-linked inheritance is synonymous with X-linked inheritance.

Males has only one X chromosome (always derived from mother) whether recessive or dominant, X-linked gene is always expressed in males. So, males are said to be *Hemizygous* (neither heterozygous nor homozygous). Father has only one X chromosome, which is always transmitted to daughters. If the trait is dominant, the daughter will be affected. If the trait is recessive, she will be a carrier for the disorder.

X chromosome cannot be transmitted from father to son, he cannot transmit X-linked trait to his male offsprings.

Most of the sex-linked diseases are associated with X chromosomes, almost all of which are recessive. The most common X-linked dominant condition is vitamin D resistant rickets.

X-linked Recessive Inheritance (Fig. 6.7 and Table 6.4)

To understand the modes of inheritance in X-linked recessive disorders, let us see the classical example of hemophilia.

Hemophilia is a disorder of blood coagulation, characterized by a tendency to excessive hemorrhage and prolonged coagulation time. A very slight trauma is sufficient to initiate hemorrhage in such patients. It is said to be due to deficiency of antihemophilic factor or globulin in the plasma. The synthesis of this factor is controlled by a gene on the X chromosome.

Table 6.4: X-linked recessive inheritance—genetic constitution of offsprings in various situations			
Sr. No.	Genes of parents	Offsprings	Results
1.	Affected male (X_hY) Normale female (XX)	X_hX, X_hX, XY, XY	Daughters—carriers Sons—normal
2.	Affected male (X_hY) Carrier female (X_hX)	X_hX_h, X_hX, X_hY, XY	Daughters—25% carriers Sons—25% normal Daughters—25% affected Sons—25% affected
3.	Normal male (XY) Carrier female (X_hX)	XX_h, XX, X_hY, XY	Daughters—25% carriers Sons—25% affected Daughter and sons—50% normal

h = hemophilia recessive gene

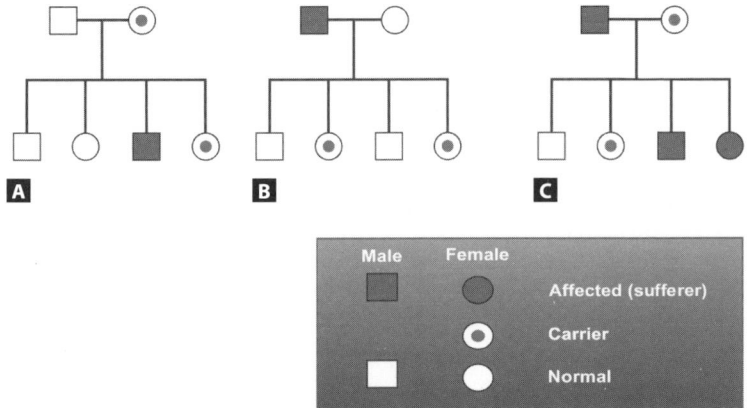

Figs 6.7A to C: Mode of X-linked recessive transmission. Note the absence of male-to-male transmission. (A) Male is normal and female is a carrier; (B) Male is sufferer and female is normal; (C) Male is a sufferer and female is a carrier

Characteristics

- An X-linked trait is usually seen in males and rarely seen in females.
- The trait is transmitted from an affected man to all his daughters and to half of his grandsons.
- There is no male to male transmission, i.e. father to son.
- The affected males in a family are related to one another through females because the trait can be transmitted through series of carrier females.

Examples

- Hemophilia
- Partial color blindness
- Duchenne muscular dystrophy
- Glucose-6-phosphate dehydrogenase deficiency
- Testicular feminization

X-linked Dominant Inheritance (Fig. 6.8 and Table 6.5)

To understand the modes of inheritance in X-linked dominant disorders, let us see the classical example of vitamin D resistant rickets.

Table 6.5: X-linked dominant inheritance—genetic constitution of offsprings in various situations			
S. No.	Genes of parents	Offsprings	Results
1.	Affected male (X_RY) Homozygous affected female (X_RX_R)	X_RX, X_RX_R, X_RY, X_RY	All children are affected
2.	Affected male (X_RY) Heterozygous affected female (X_RX)	X_RX_R, X_RX, X_RY, XY	All daughters are affected and half of the sons are normal and half are affected
3.	Normal male (XY) Homozygous affected female (X_RX_R)	XX_R, XX_R, X_RY, X_RY	All children are affected
4.	Affected male (X_RY) Normal female (XX)	X_RX, X_RX, XY, XY	All daughters are affected and all sons are normal
5.	Normal male (XY) Heterozygous affected female (X_RX)	XX_R, XX, X_RY, XY	Half of the sons and daughters are affected and half are normal

R = affected gene

78 Basic Human Genetics

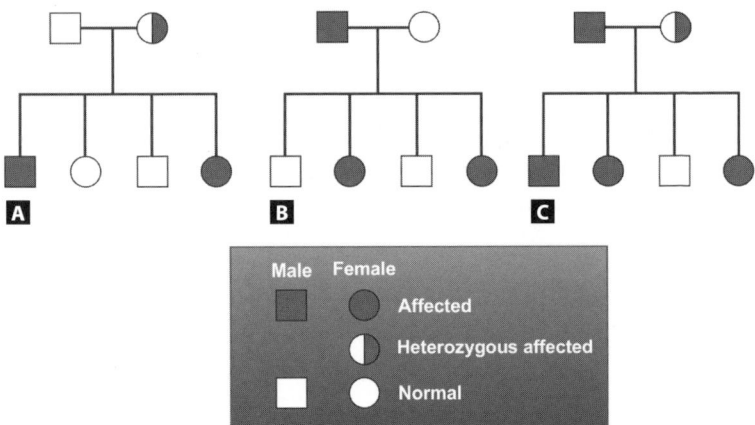

Figs 6.8A to C: X-linked dominant transmission. Only females are affected. Usually males who inherit the mutant allele die in utero. (A) Normal male and female Heterozygous affected (sufferer); (B) Affected male and normal female. (C) Male affected and female heterozygous affected

Characteristics

- Affected females are twice as common as affected males.
- Affected male transmit the trait to all his daughters, but none of his sons.
- Affected females pass the trait to both sons and daughters.

Examples

- Xg blood group
- Vitamin D resistant rickets
- Hypophosphatemia.

Y-linked Inheritance

This is also known as holandric inheritance. The most common known traits are hairy pinna and baldness (Fig. 6.9).

As the Y chromosome is only found in males, Y linked genes are only found in males. Males alone get affected and affected males transmit the trait to all his sons but to none of his daughters. As the Y chromosome is single, genes on Y chromosome are always expressed phenotypically. Recent studies suggest that H-Y histocompatibility antigen and genes responsible for spermatogenesis are located on Y chromosome.

Characteristics

- An X-linked trait is usually seen in males and rarely seen in females.
- The trait is transmitted from an affected man to all his daughters and to half of his grandsons.
- There is no male to male transmission, i.e. father to son.
- The affected males in a family are related to one another through females because the trait can be transmitted through series of carrier females.

Examples

- Hemophilia
- Partial color blindness
- Duchenne muscular dystrophy
- Glucose-6-phosphate dehydrogenase deficiency
- Testicular feminization

X-linked Dominant Inheritance (Fig. 6.8 and Table 6.5)

To understand the modes of inheritance in X-linked dominant disorders, let us see the classical example of vitamin D resistant rickets.

Table 6.5: X-linked dominant inheritance—genetic constitution of offsprings in various situations

S. No.	Genes of parents	Offsprings	Results
1.	Affected male (X_RY) Homozygous affected female (X_RX_R)	X_RX, X_RX_R, X_RY, X_RY	All children are affected
2.	Affected male (X_RY) Heterozygous affected female (X_RX)	X_RX_R, X_RX, X_RY, XY	All daughters are affected and half of the sons are normal and half are affected
3.	Normal male (XY) Homozygous affected female (X_RX_R)	XX_R, XX_R, X_RY, X_RY	All children are affected
4.	Affected male (X_RY) Normal female (XX)	X_RX, X_RX, XY, XY	All daughters are affected and all sons are normal
5.	Normal male (XY) Heterozygous affected female (X_RX)	XX_R, XX, X_RY, XY	Half of the sons and daughters are affected and half are normal

R = affected gene

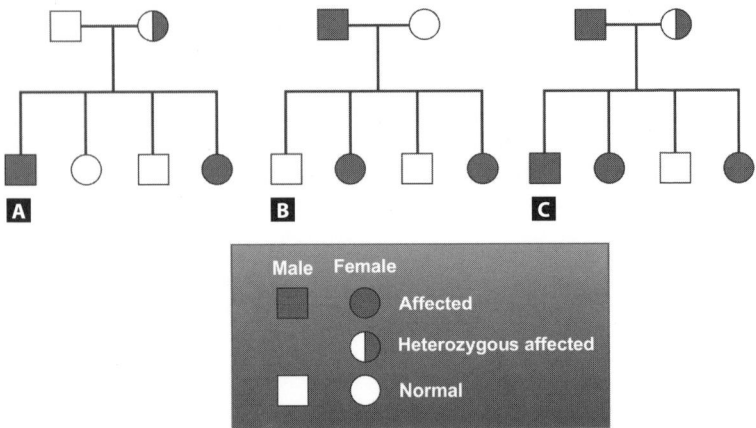

Figs 6.8A to C: X-linked dominant transmission. Only females are affected. Usually males who inherit the mutant allele die in utero. (A) Normal male and female Heterozygous affected (sufferer); (B) Affected male and normal female. (C) Male affected and female heterozygous affected

Characteristics

- Affected females are twice as common as affected males.
- Affected male transmit the trait to all his daughters, but none of his sons.
- Affected females pass the trait to both sons and daughters.

Examples

- Xg blood group
- Vitamin D resistant rickets
- Hypophosphatemia.

Y-linked Inheritance

This is also known as holandric inheritance. The most common known traits are hairy pinna and baldness (Fig. 6.9).

As the Y chromosome is only found in males, Y linked genes are only found in males. Males alone get affected and affected males transmit the trait to all his sons but to none of his daughters. As the Y chromosome is single, genes on Y chromosome are always expressed phenotypically. Recent studies suggest that H-Y histocompatibility antigen and genes responsible for spermatogenesis are located on Y chromosome.

Inheritance 79

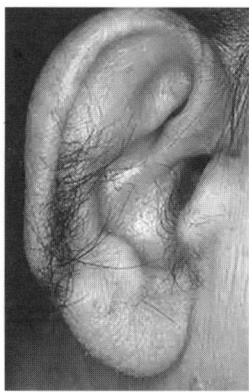

Fig. 6.9: Hairy pinna, a condition determined by a gene linked with Y chromosome

Codominant Inheritance

When both the allelic genes are dominant but of two different types, both traits may have concurrent expression in heterozygote state, such types of genes and the traits are said to be codominant. In ABO blood groups, A gene and B gene both are dominant; when they occupy identical loci in homologous chromosomes, AB blood group is expressed.

Intermediate Inheritance

Certain characters do not follow simple dominant or recessive kind of inheritance. In such cases, a heterozygote resembles neither a homozygote with both dominant genes nor one with both recessive genes. Such characters exhibit intermediate inheritance.

This is very well exemplified by sickle cell anemia and sickle cell trait. The homozygotes for the abnormal allele has severe sickle cell anemia. The heterozygote for the abnormal allele does not have severe sickle cell anemia nor is he completely normal. A proportion of his red cells show the sickling phenomenon. Such a heterozygote is intermediate between normal homozygotes and sickle cell homozygotes; and is said to have sickle cell trait.

Both these conditions are characterized by the presence of an abnormal hemoglobin.

Basically, normal hemoglobin is a protein containing amino acids arranged in the form of 2 alpha chains and 2 beta chains. The abnormal

80 Basic Human Genetics

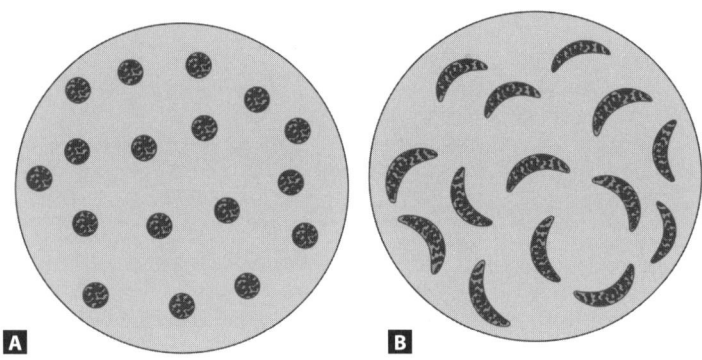

Figs 6.10A and B: (A) Normal erythrocytes; (B) Sickled erythrocytes

hemoglobin of sickled red blood cells differs from normal hemoglobin in presence of valine replacing glutamic acid of beta chain.

The abnormal hemoglobin results in lower concentration of oxygen. Consequently red blood cells become sickle-shaped and may rupture, causing hemolytic anemia (Figs 6.10A and B). The two conditions are compared in Table 6.6.

Table 6.6: Comparison of sickle cell trait and sickle cell anemia		
	Sickle cell trait	*Sickle cell anemia*
Occurrence	It is seen in a heterozygote with one normal and one abnormal gene	It is seen in a homozygote with both abnormal genes
Amount of abnormal hemoglobin	30%	100%
Severity	Less severe (Detected only by examination of blood)	Fatal

Polygenic Inheritance

Some inherited traits instead of being governed by single gene, (allele at one locus) are determined by number of genes, each having minor effect in expression of a single trait. Such traits are known as polygenic traits and the pattern of inheritance is known as polygenic inheritance.

Characters such as height, weight or intelligence do not follow the simple Mendelian patterns of inheritance. Such characters may be controlled by several genes located at different loci, having similar effects.

Certain conditions, not only follow the polygenic inheritance, but are also influenced by the environment. This type of inheritance is known as M*ultifactorial inheritance*. Examples of such conditions are peptic ulcer, essential hypertension, cleft lip, cleft palate and diabetes mellitus.

Mitochondrial Inheritance

The great majority of genetic diseases are caused by defects in the nuclear genome, however a small but significant number of diseases can be caused by mutations in mitochondrial DNA.

Each human cell contains several hundreds or more mitochondria in its cytoplasm. Mitochondria is responsible for generation of ATP in the body, which is the main source of energy for all metabolic activities.

The mitochondria have their own DNA molecules mtDNA, that contain genes. Because the sperm contains hardly any cytoplasm, the mitochondria in the zygote originate almost exclusively from cytoplasm of the ovum. Therefore, mitochondrial inheritance of a trait is exclusively maternal, inherited by all offsprings with males and females being equally affected.

Examples of mitochondrial inheritance are neurogenic muscle weakness, mitochondrial encephalomyelopathy and hereditary optic neuropathy, etc.

7
Population Genetics

■ INTRODUCTION
Population genetics is the study of genes in the populations.

Objectives
Its objects are to study:
- Genes and gene frequencies in various populations.
- Distribution of inheritance of genes and traits in the populations.
- Mechanisms involved in evolution of populations.

The studies are based on mathematical calculations, environmental factors and population migration. It helps to study etiology and course of some heritable disorders and for calculations of autosomal recessive gene carrier frequencies.

Hardy-Weinberg Principle
G.H. Hardy, a mathematician and W. Weinberg a physician defined this law in 1908. The law states that the frequency of alleles for any character will remain unchanged in a population through any number of generations, unless this frequency is altered by some outside influence such as mutation, population migration, nonrandom mating and selection.

Gene Pool
It is defined as the sum total of genes of all the individuals of a given population.

■ DERMATOGLYPHICS
Introduction
- The word dermatoglyphics comes from two Greek words (derma = skin, glyphe = carve).

- It involves the study of dermal ridge configurations on the digits, palms and soles.
- The term was coined by Dr Harold Cummins.

Development

The dermal ridges begin to develop around 13th week of fetal life and the pattern formation is completed by the 19th week.

Different Patterns

Various patterns studied under the science of dermatoglyphic are:
- The flexion creases of palm
- Dermal patterns
  - Finger patterns
  - Palmar patterns
  - Plantar patterns.
- The flexion creases of palm (Fig. 7.3)
  Flexion creases in palm are known as head, heart and life line in the field of palmistry. In Down's syndrome, patients show a single flexion crease in the palm known as *Simian crease.* Such single transverse flexion crease is also found in Trisomy 13, Trisomy 18 and in cri du chat syndrome. It may also be found in 1% of normal subjects
- Dermal patterns.

Finger Patterns (Fig. 7.1)

The patterns on the fingertip can be classified according to Galton system into 3 main types, on the basis of number of triradii present. Triradius is a point from which three ridges course in three different directions at angles of about 120°.
- *A simple arch:* It has no triradius
- *A loop:* It has one triradius which can be ulnar or radial depending on the side to which it opens
- *A whorl:* It has two or more triradii.

The size of a finger pattern is considered as the ridge count. It is the total number of ridges come across a line drawn from the triradius to the core (center) of the finger pattern. The arch has zero ridge count because it has no triradius. The total ridge count (TRC) means sum of ridge counts of 10 fingers. High TRC is found in Turner's syndrome and low TRC is found in Klinefelter's syndrome. It is studied that greater the number of sex chromosomes the lower the TRC.

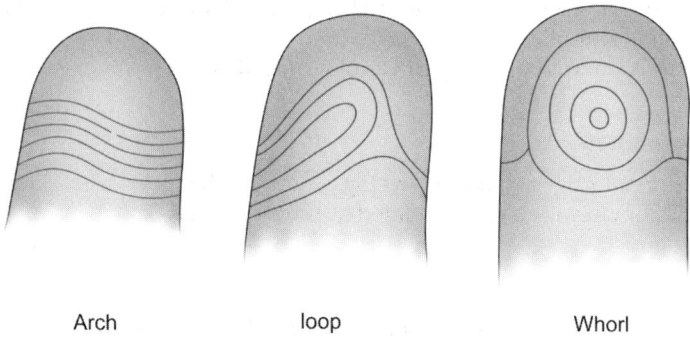

Arch loop Whorl

Fig. 7.1: Finger patterns

The most frequently seen finger pattern is an ulnar loop. Whorls are mostly found on 1st, 2nd and 4th digits. Arches and radial loops are most infrequently seen.

Variation in the pattern frequencies is seen on two sides and also in two sexes. Females have slightly more arches and fewer whorls than males. Some racial differences also exist in pattern frequencies.

Palmar Patterns (Fig. 7.2)

The normal palm has a triradius commonly placed over the fourth metacarpal near the base of the palm between the thenar and hypothenar areas. This is called the axial triradius (t). Another four digital triradii a, b, c, and d are present near the distal border of the palm. Usually, the 'atd' angle measures around 60°. In Down's syndrome, the axial triradius is shifted distally and so, 'atd' angle is greater than 75°.

Plantar Patterns

They are present on the sole of the foot. Very little is known about the plantar pattern. About 50% of Down's syndrome patients show an unusual "arch tibial" pattern.

Significance of Dermatoglyphics

- Characteristic patterns are seen in various chromosomal disorders, e.g. Down's syndrome.
- Such studies are very helpful in determination of twin zygosity.
- The study is applied in medicolegal cases like identification of a suspected criminal.

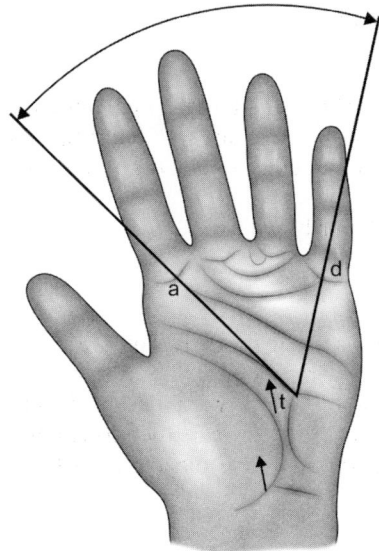

Fig. 7.2: Palmar patterns showing atd angle

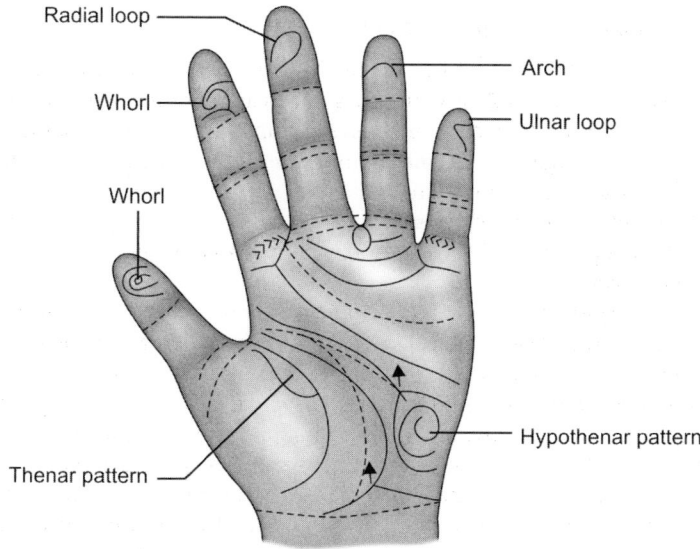

Fig. 7.3: The nomenclature of the dermatoglyphics of the palm and finger

■ STUDY OF TWINS

Introduction
Birth of two young individuals in single pregnancy, is known as twinning. Twinning in humans is a hereditary trait. The importance of twin studies to assess the effects of heredity and environment was stressed by Galton in 1874.
Incidence: 1 in 80–90 births.

Basic Concept
If genetic factors play a role in the causation of a disease, it will affect both members of a twin, more frequently in monozygotic than in dizygotic twins. If only one of the members of a monozygotic twin is expressing a particular trait/disorder, it can be concluded that nongenetic (environmental) factors are also playing a part in the etiology of the disorder.

Types of Twinning
- Monozygotic twinning
- Dizygotic twinning.

Monozygotic Twinning (Figs 7.4A to C)
Two embryos are derived from a single ovum, which is fertilized by a single sperm. Monozygotic twins are same in appearance, structure, sex, fingerprints, blood groups and genetic constitution. Transplantation of tissues and organs from one member of identical twins to the other member is accepted without rejection.

Depending on the stage of development at which zygote splits to form two separate embryos, monozygotic twins are classified under following types:
- *Monozygotic bichorionic (25–30%)*: At two-cell stage of cleavage division, the blastomere cells develop into two separate zygotes. Both the members possess separate placenta and separate chorionic sac.
- *Monochorionic biamniotic (70–75%)*: At the blastocyst stage, the inner cell mass is divided in two equal parts and each develops into separate embryo. Each embryo is invested by a separate amniotic membrane and having a single placenta and single chorionic sac.
- *Monochorionic monoamniotic (1–2%)*: If division of inner cell mass occurs after formation of amniotic cavity, i.e. after 8 days of fertilization then the monozygotic twins shall have one amnion and one chorion.

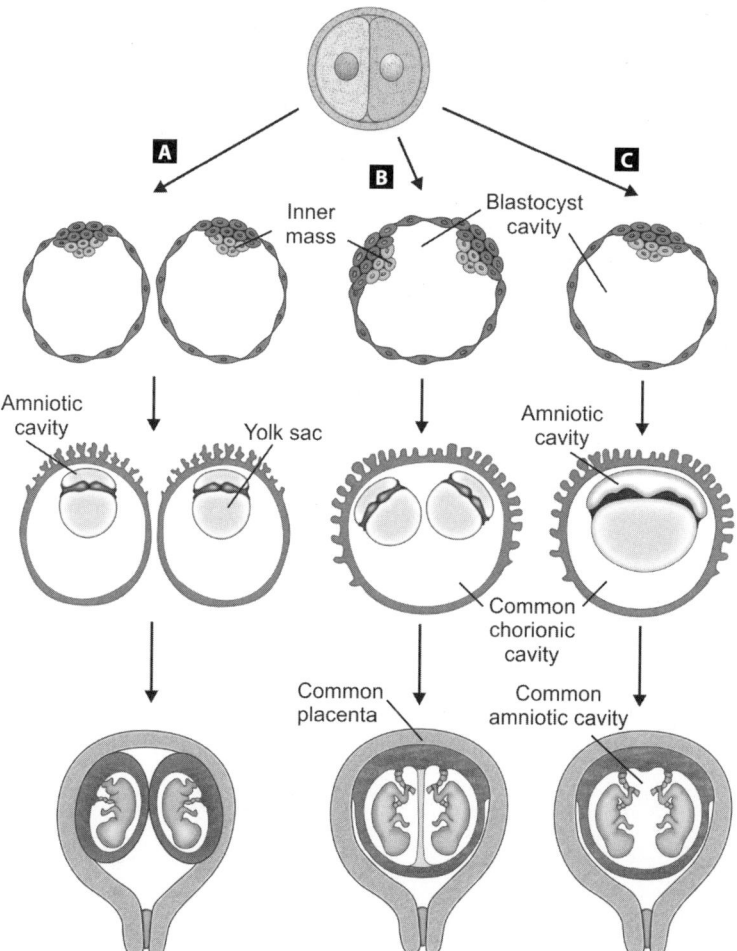

Figs 7.4A to C: Monozygotic twinning: (A) Monozygotic bichorionic; (B) Monochorionic biamniotic; (C) Monochorionic monoamniotic

Dizygotic Twinning (Fig. 7.5)

Two embryos develop by two ova discharged in a single ovarian cycle and fertilized by two different sperms. The resulting twins are fraternal or unlike conveying different genetic material. They may be of same or

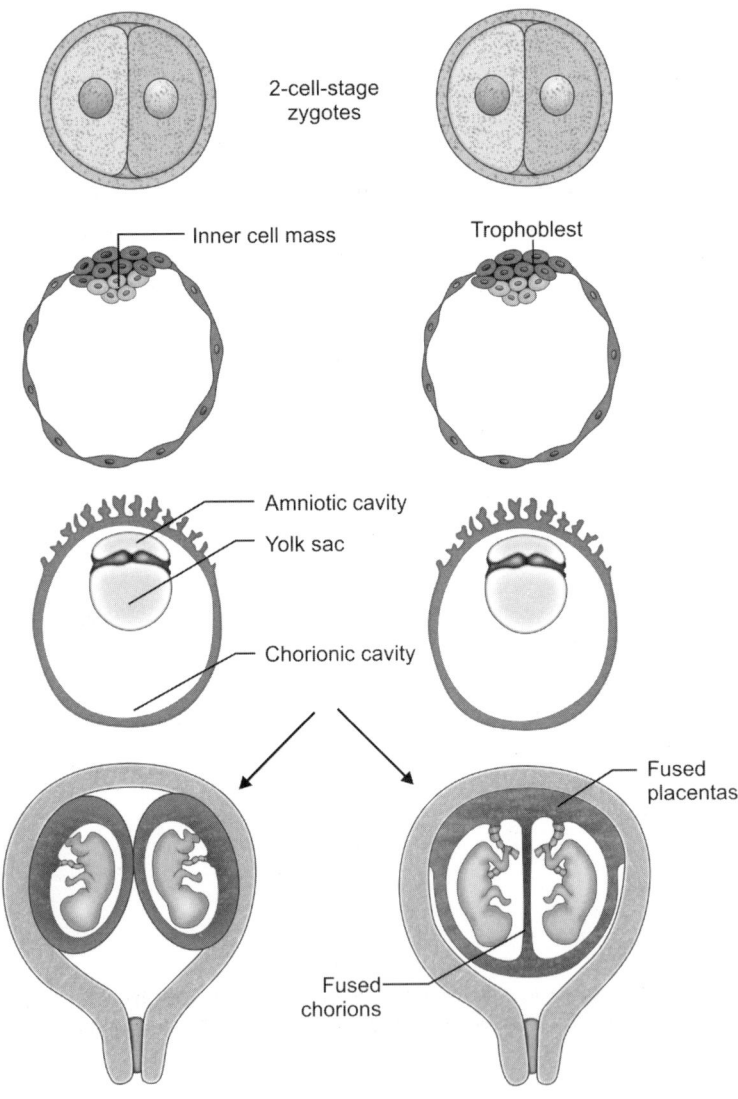

Figs 7.5: Dizygotic twinning

different sex. They possess separate placentae and separate chorionic sacs.

Abnormal twinning
- Conjoined twins
- Chimera.

Conjoined Twins
When inner cell mass fails to separate completely, it results in development of two embryos who are incompletely separated from each other, known as conjoined twins or double monster.

Chimera
A chimera is a person whose cells are derived from more than one fertilized egg.

In some cases of dizygotic twinning, there is exchange of genetically dissimilar cells through the placental blood vessels. For example, suppose one twin is male having blood group A and his co-twin is a female having blood group B. In such condition, it is not uncommon that the male twin may possess few red blood cells having blood group B and female possess few red cells having blood group A.

Tests of Zygosity
Such tests are useful while planning for surgery, e.g. kidney transplant

The zygosity can be determined by:
- Blood group studies, e.g. ABO, Rh
- Histocompatibility antigens
- Serum proteins
- Appearance of placenta, amnion and chorion.

Application of Twin Studies
- Study of identical twins under different conditions is one of the methods of distinguishing the postnatal influences of heredity and environment
- Such studies are also useful in exploring the possibility of genetic factors in the causation of diseases
- In transplantation, a monozygotic co-twin is the most useful donor, whereas a dizygotic co-twin is genetically like any other sibling.

8
Prevention and Treatment of Genetic Diseases

■ PREVENTION OF GENETIC DISEASES

Genetic Counseling

Introduction

Genetic counseling plays a vital role in prevention of genetic disorders. A genetic counselor functions to inform the individuals about nature of present and future possible genetic diseases. He helps the individual concerned in deciding the line of action.

Who Can be a Genetic Counselor?

The term genetic counselor refers to any medical professional who is professionally qualified to provide genetic counseling. Typically, a genetic counselor is a genetics professional with a master's degree or PhD.

A Good Counselor should have the Following Qualities

- He should have strong knowledge of principles of genetics.
- He should be aware of diseases of genetic origin.
- He should be tactful and have a sympathetic and kind approach.

Genetic counseling may be given prospectively or retrospectively.
- *Prospective genetic counseling*: It involves identification of heterozygous individuals by various screening procedures. Such individuals are also made aware of the risk of their children getting the disease if they marry another heterozygote for the same gene. Presently, this approach is confined to populations in which frequency of a specific disease is high, e.g. sickle cell disease in West African population.
- *Retrospective genetic counseling*: Mostly, genetic counseling is done retrospectively. Different methods can be employed for this purpose, such as sterilization and termination of pregnancy.

Role of Genetic Counselor

- Genetic counselor can predict the characteristics of the future generation progeny, thereby helping in planning the parenthood.
- A genetic counselor can assess the probability of having a child with a specific hereditary defect. Karyotype analysis of parents can provide clues regarding any chromosomal abnormality which can be transmitted to subsequent generations.
- A counselor can create awareness regarding role of consanguineous marriages in producing a defective baby. Such marriages carry an increased risk for the child in getting traits controlled by recessive genes, e.g. phenylketonuria.
- Genetic counselor should stress that late marriages should be discouraged. It is well-established that Down's syndrome is more frequently seen in children born of elderly mothers.
- A counselor should make people aware of hazards of radiations.
  - Patients undergoing radiological examination should be protected against unnecessary exposure of gonads to radiations.
  - Radiological examination of pregnant women should be carried out only when it is indispensable.
- A genetic counselor can predict the occurrence of sex-linked disorders, by conducting examination of cells from amniotic fluid.
- A counselor can render advice regarding the importance of voluntary abortions in the prevention of suspected metabolic and hereditary disorders.
- A genetic counselor should also highlight the hazards of certain drugs, if not taken with care. For example, certain antimalarial drugs can precipitate hemolysis in persons with genetic deficiency of Glucose-6-phosphate dehydrogenase.

Prenatal Diagnosis

Indications for prenatal diagnosis:
- Advanced maternal age
- A previous baby with a chromosomal abnormality
- Determination of fetal sex in couples at risk for a disease inherited in X-linked recessive manner
- Parents with any chromosomal abnormality
- Family history with some chromosomal abnormality.

Congenital defects can be defined as significant, definable, structural and/or developmental abnormalities observed at birth. Some anomalies are due to environmental agents, e.g. viral infections of mother and ionizing radiation *in utero*. Certain abnormalities are because of mutant gene or chromosomal aberrations, while majority of them are multifactorial in origin.

Until now, there is no single screening test is available to detect all congenital abnormalities in all pregnant women. The tests available are specific for particular types of abnormalities.

The most common cause of perinatal mortality and morbidity is due to various congenital malformations. As we know that genetic disorders are mostly nontreatable, one should think about preventive steps for it. Therefore, prenatal diagnosis plays an important role to identify specific genetic risk and to refer the patients for expert genetic counseling and further tests.

The methods available for prenatal diagnosis are following:
- Noninvasive
    - Ultrasound
    - Fetal echocardiograms
    - Computerized tomography (CT) and magnetic resonance imaging (MRI)
- Invasive
    - Serum alpha-fetoprotein (AFP)
    - Amniocentesis
    - Chorionic villous sampling
    - Fetal blood sampling
    - Fetoscopy.

■ NONINVASIVE TECHNIQUES

Ultrasonography

It is the most standard and safe method for fetal evaluation. It is useful for:
- Determination of fetal age, growth and development
- To ascertain multiple gestation
- To locate placental implantation site and calcification
- To detect gross fetal malformation such as anencephaly
- To detect chromosomal abnormality by measuring *nuchal translucency*. Nuchal translucency (NT) is the area just under the skin at the back of the fetal neck, which can be measured at 10–14 weeks of gestation. Increased NT thickness is associated with trisomy 13, 18, 21 and Turner's syndrome.

Fetal Echocardiogram

This procedure is performed during 20-23 weeks through maternal abdomen to rule out congenital cardiac defects in high-risk pregnancy. It provides information regarding size, position, valves and chambers of the heart of the fetus.

Computerized Tomography and MRI

It can be used during pregnancy, but there may be long-term side effects particularly with MRI due to the high noise level transferred to the fetus.

■ INVASIVE TECHNIQUES

Serum Alpha-fetoprotein Level (AFP)

Maternal serum AFP screening is used routinely to assess for neural tube defect. Screening is usually performed between 16-18 weeks of gestation.

Raised AFP level is found in following conditions:
- Neural tube defects (NTD)
- Multiple gestation
- Rh incompatibility
- Low gestational age
- Missed or threatened abortions
- Intrauterine death.

Decreased AFP level is found in following conditions:
- Trisomy 13, 18 and 21.

Triple marker test includes measurement of serum AFP, serum human chorionic gonadotropin and serum estriol level. This study is especially useful for testing trisomies.

Amniocentesis

Tapping of amniotic fluid per abdomen in a pregnant woman is known as amniocentesis (Fig. 8.1).

Procedure

The procedure is performed between 16 and 20 weeks when sufficient amount of amniotic fluid is available. The needle is inserted under ultrasound guidance per abdomen to collect the amniotic fluid. Care is taken not to penetrate the placenta and 10-20 cc of clear amniotic fluid is collected. The collected sample of amniotic fluid is subjected to

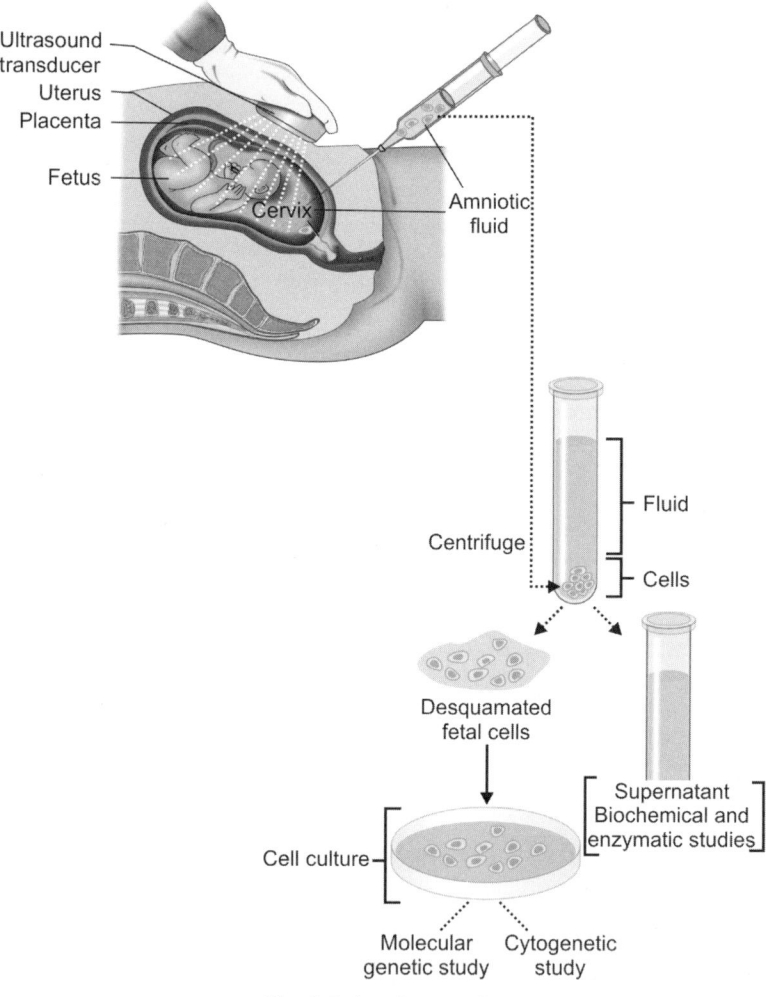

Fig. 8.1: Amniocentesis

laboratory investigations. Cells are separated from amniotic fluid and studied to obtain a fetal karyotype. Result takes about 10–28 days. The fluid component is sent for biochemical analysis. Raised level of alpha fetoprotein indicates neural tube defect.

Risk
- Spontaneous abortion risk in 1–2%
- Nausea and abdominal pain

- Placental or umbilical cord perforation
- Premature labor
- Rh sensitization.

Chorion villus biopsy (CVB)

CVB is performed between 8 and 12 weeks. It provides earliest diagnosis of fetal problem. Under ultrasound guidance, the procedure is done by aspirating chorionic villi either transabdominally or tanscervically depending on placental position (Fig. 8.2).

Advantages

- It provides a diagnosis at a much earlier stage (first trimester) than amniocentesis (second trimester). So if medical termination of pregnancy is needed it can be done safely
- Direct examination of fetal cell is possible.

Disadvantages

- It carries a greater risk of abortion than amniocentesis
- Alpha fetoprotein level cannot be measured, so the neural tube defect cannot be diagnosed.

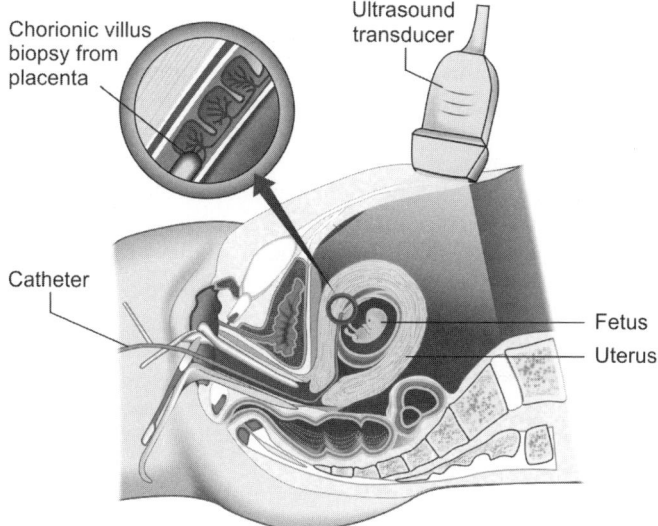

Fig. 8.2 : Chorion villous biopsy

Fetal blood Sampling (Percutaneous Umbilical Blood Sampling)

It means collection of blood from the fetal-placental circulation. It is performed between 16 and 18 weeks of gestation and results can be obtained in 2-3 days. A needle is passed through maternal abdomen and blood is collected from umbilical blood vessels under ultrasound guidance.

This procedure is needed for making prenatal diagnosis in various conditions like hematological diseases and immunological disorders. It is advantageous over CVB and amniocentesis as it provides rapid diagnosis of a disorder but chances of fetal loss is more with fetal blood sampling.

Fetoscopy

This procedure involves introduction of an illuminated instrument into the uterus to visualize the fetus. It can help in detecting cleft lip, extra fingers and neural tube defects. Also, the operator can obtain a sample of fetal blood through the fetoscope to diagnose blood disorders like hemophilia and thalassemia.

Nowadays, because of availability of more advanced technique, usually it is not employed.

■ TREATMENT OF GENETIC DISEASES

All genetic diseases cannot be treated. The principal approach to the control of the genetic disease is therefore, prevention through genetic counseling, with prenatal diagnosis and selective abortion where possible. But, of course, situation is not as dark, as it may appear to be. Various forms of treatment are available.

- *Elimination in diet*: Elimination of certain substances in diet like
  - Phenylalanine in phenylketonuria,
  - Reduction of diet with high cholesterol in familial hypercholesterolemia
  - Carbohydrate in galactosemia.
- *Supplementation in diet*: Vitamin C supplementation in deficiency of vitamin C synthesis.
- *Replacement therapy*:
  - Patients of hemophilia can be helped by administering antihemophilic globulin.
  - Replacement of vitamin D in vitamin D resistant rickets.
  - Thyroxin in congenital hypothyroidism

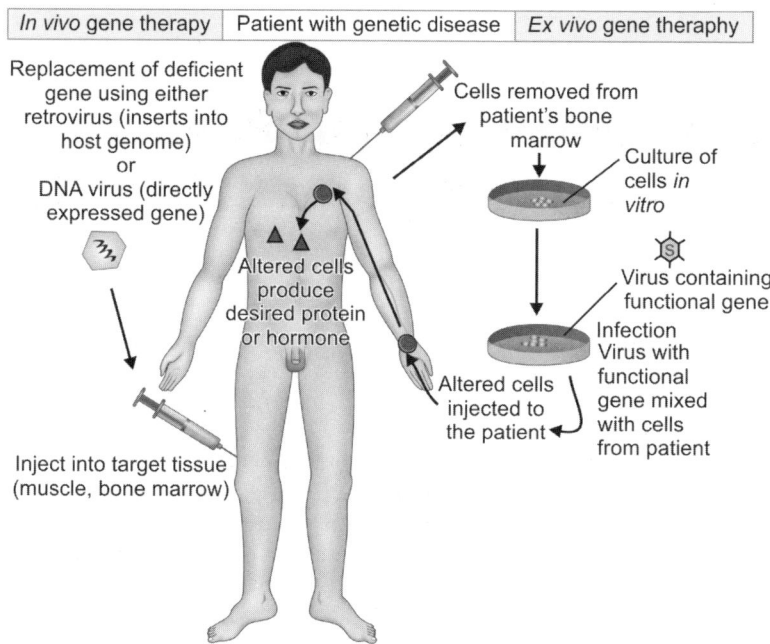

Fig. 8.3: Gene therapy. *In vivo* gene therapy delivers genetically modified cells directly to the patient. *Ex vivo* gene therapy removes cells from the patient which are modified in vitro and then returned to the patient

- *Surgical procedures*:
    - Spina Bifida can be corrected by surgical methods
    - Colectomy can be done in polyposis coli
    - Organ replacement includes kidney transplant in adult polycystic disease
- *Gene therapy (Fig. 8.3)*: The term gene therapy means to treat a disorder by modifying the genetic material of living cells. Gene therapy includes:
    - Replacement of a mutant gene causing the disease with a healthy copy of the gene
    - Inactivation of the mutant gene, responsible for abnormal functioning
    - Introduction of a new gene into the body which help to fight the disease.

Though gene therapy is used as one of the treatment modalities for treating number of disorders like inherited diseases, some type of cancers, etc. there is a doubt regarding its safety and effectivity.

9
The Human Genome Project and Recent Advances in Genetics

A genome is an organism's complete set of DNA, including all of its genes. The human genome project was an international research effort to determine the sequence of the human genome and identify the genes that it contains.

It is one of the most ambitious project in the field of biomedical research. The project was coordinated by the national institutes of health and the US Department of Energy. Additional contributors were United Kingdom, France, Germany, Japan, China and India. The project began in 1990 and was completed in 2003 two years ahead of its original schedule.

■ GOALS

- To provide a genetic marker map
- To provide a physical map
- To provide the complete 3 billion bp sequence of the human genome.

The *marker map* was completed early in the course of the project and includes variable sequences (polymorphisms), situated across the entire genome serve as genetic markers. Various polymorphisms include restriction fragment length polymorphisms (RFLPs), variable number of tandem repeats (VNTRs) and short tandem repeat polymorphism (STRPs). The identification of the genetic markers constitutes the genetic mapping of the genome. A closely linked marker can be found for virtually any disease causing gene.

The *physical map* includes cutting up the genome into smaller segments, cloning them into vectors, and characterizing the sequence of their ends to get small sequence tags (sequence tagged sites, STS) across the genome.

The final goal, the complete human genome has been achieved in 2003. The project identified the locations of many human genes and provided information about their structure and organization.

After genomic mapping it became possible to assemble randomly sequenced DNA segments into larger sequences and subsequently to assemble the entire human genome. Genomic DNA was cut up into smaller bits using restriction enzymes and cloned into vectors. Each of these small segments was sequenced with the help of software programs to overlap in order to assemble the larger DNA sequence to make a complete human genome. A complete human genomic sequence provides the ultimate genetic blueprint of the human species.

In addition, the Human Genome Project, sequenced the genomes of several other organisms like medically significant viruses and bacteria, agriculturally important crops such as rice and maize and important experimental organisms such as yeast, fruit flies, mice, etc. Similarities between the genes of these organisms and humans have helped us to understand the nature of many human genes.

■ BENEFITS OF HUMAN GENOME PROJECT

- Previously, it took several years to identify mutation within genes which are responsible for inherited diseases. Recently with the availability of genomic sequence and the comprehensive mapping of the genome to identify a mutant gene became easier.
- The vast number of genes and their protein products are considered potential drug targets for the treatment. With the help of the genome project the potential for manufacture of gene products by recombinant DNA techniques has increased.
- At last, completion of the project improved the treatment through more specific drugs or gene therapy.

■ RECENT ADVANCES IN GENETICS

Gene Libraries

A genomic library is a collection of the total genomic DNA from a single organism.

There are different types of DNA libraries:
- cDNA libraries—formed from reverse-transcribed RNA.
- Genomic libraries—formed from genomic DNA.

Steps for creating a genomic library (Fig. 9.1):
- DNA is extracted from a cell of an organism.
- DNA is digested with a restriction enzyme which cut the DNA into small fragments of specific size each containing one or more genes.

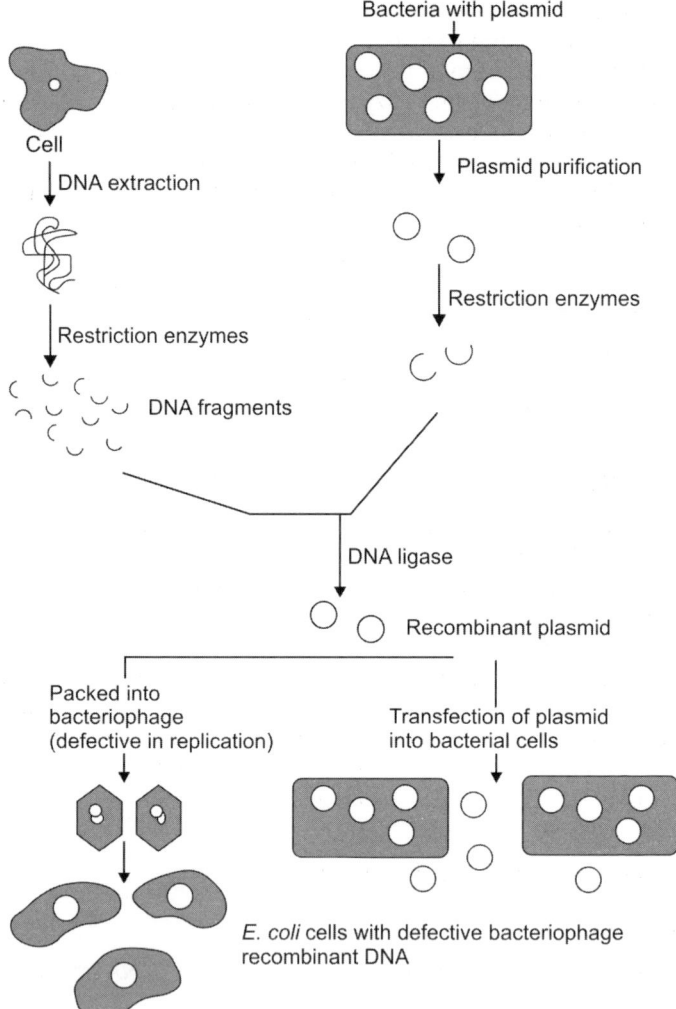

Fig. 9.1: Steps for creating a genomic library

- Fragments of DNA are inserted into vectors, the enzyme DNA ligase is used to seal the DNA fragments into the vector. This creates a large pool of recombinant molecules.
- The newly formed recombinant molecule is then introduced into host cells where it can replicate. The host cell is then cloned in order to produce enough material to create a Genomic library.

Applications of Genomic library
- Determining the complete genome sequence of a given organism.
- Helps to understand the molecular basis of various diseases.
- Human proteins like insulin and growth hormone can be produced
- Study of genetic mutations in cancer tissues.

Gene mapping, also called genome mapping, is the creation of a genetic map assigning DNA fragments to chromosomes.

Index

Page numbers followed by *f* refer to figure, *t* refer to table

A

Abdomen 45, 50
Abortions
　missed 93
　spontaneous 94
　threatened 93
Achondroplasia 71
Acridine 64
Alleles 69
Amenorrhea 48
Amniocentesis 92, 93, 94*f*, 95
Amniotic fluid cells 11
Angelman syndrome 37
Anonychia 72
Aorta, coarctation of 50
Aortic stenosis 50
Atresia, duodenal 45
Autosomal dominant 74*t*
　inheritance 70, 71, 71*f*, 72*t*, 74
Autosomal recessive 74*t*
　inheritance 70, 72, 74
Autosomes 10, 42
Axillary hair 50
Azospermia 53

B

Barr bodies 10, 16, 53
Bone marrow 11
Brachydactyly 72
Brain malformation 45
Breasts, underdeveloped 50

C

Caffeine 64
Cancer, breast 53
Carcinogens, chemical 35
Cardiovascular system 50
Cell cycle 22, 23
　stages in 23*f*
Cell division 21
　abnormal 33
　types of 25
Cell metabolism, study of 64
Centric fusion 38
Centromere
　number of 8, 9
　position of 8, 9
Centromeric index 13
Centromeric probe 16
Chimera 89
Chorionic villus
　biopsy 95, 95*f*
　sampling 92
Chromatin 24
　condensed 24
　extended 24, 24*t*
　sex 16
Chromosomal aberrations 35
Chromosomal abnormalities 35
Chromosomal analysis 50
Chromosomal complement 53
Chromosomal mutation 61

Chromosome 6
  complement 51
  length of 13
  philadelphia 42, 47
  ring 36, 37, 37*f*
  sex 7, 10, 35
  shape of 13
  structure of 7*f*
  types of 9*f*
Color blindness, partial 77
Congenital heart disease 46
Congenital hypothyroidism,
    thyroxin in 96
Cri du chat syndrome 36, 42, 42*f*
Cubitus valgus 50
Cushing's syndrome 50
Cystic fibrosis 73
Cytokinesis 21

D

Daughter cells 32*f*
Denver system 8, 10, 13
Deoxyribonucleic acid 8, 60
Diakinesis 29
Diploid number 40
Down's syndrome 19, 35, 42, 43, 44*f*, 83, 84
Duchenne muscular dystrophy 77

E

Edward's syndrome 42, 46, 47*f*
Epicanthal folds 44, 50
Erythrocytes, normal 80*f*

F

Fetal blood 11
  sampling 92, 96
Fetal bovine serum 12
Fetal echocardiograms 92
Fetoscopy 92, 96
Fluorescent bodies, study of 10, 19
Fluorescent in situ hybridization
    (FISH) 10, 15, 15*f*
  types of 16
  uses of 16

Follicle-stimulating hormone 51, 52
Formaldehyde 64
Fragile X syndrome 48, 54

G

Genes 56
  chemical basis of 58
  classification 56
  definition 56
  pool 82
  therapy 97, 97*f*
Genetics 1, 2, 99
  biochemical 3
  counseling 90
  developmental 3
  diseases
    prevention of 90
    treatment of 90, 96
  molecular 3
  population 3, 82
  radiation 3
Germ cells 33
Gestation, multiple 93
Glucose-6-phosphate dehydrogenase
    deficiency 77
Gonadal dysgenesis 50
Growth retardation 42
Gynecomastia 51

H

Hair, scanty growth of 51
Haploid number 40
Hardy-Weinberg principle 82
Heart defects 45
Hemophilia 77
Hernia, umbilical 45
Histones 8
Homozygous 69
Human genome project 99
  benefits of 99
Human male karyotype 14*f*
Hybridization 15
Hyperploidy 41
Hypophosphatemia 78

Hypoplasia 45, 50
Hypotonia 43

I

Incubation 12
Infertility 53
Intersex 48, 54
Intrauterine death 93
Inversion 36, 39
Isochromosome 36

K

Karyokinesis 21
Karyotype 10
  preparation of 13
Klinefelter's syndrome 17, 19, 35, 48, 51, 52f, 83

L

Low gestational age 93
Luteinizing hormone 52
Lymphedema, peripheral 50
Lyon's hypothesis 18

M

Magnetic resonance imaging 92
Meiosis 25, 28, 29, 32, 33t
  normal 32f
  significance of 32
Meiotic division 34f
Mendel's laws of inheritance 66, 67
Mental retardation 43
Mental status 43
Messenger RNA 60
Metacarpal sign, positive 50
Micrognathia 43, 45, 46
Mitosis 25, 26, 33t
  stages of 27f
Mitotic figures 28
Monosomy 41
Mutation 60
  deletion 61, 61f
  duplication 61
  frameshift 61, 62f

  germ cell 63
  insertion 61, 62f
  missense 62, 63f
  nonsense 62, 63f
  somatic cell 63
Myeloid leukemia 48f

N

Nails, dystrophic 50
Nausea 94
Neonatal calf serum 12
Neural tube defects 93
Nuchal translucency 92
Nucleic acid 60
  basic structure of 59f

O

Osteogenesis imperfecta 71
Ovaries, absence of 50

P

Pachytene 29
Pain, abdominal 94
Palmar patterns 83, 84
Paracentric inversion 40
Patau's modification 13
Patau's syndrome 42, 45, 46f
Pedigree analysis 69
Pedigree chart, symbols used in 70f
Pelvis 45, 50
Percutaneous umbilical blood
  sampling 96
Phenol 64
Phenylketonuria 66, 73
Plantar patterns 83, 84
Polydactyly 45
Polymorphonuclear cell 17f
Polymorphonuclear leukocytes 10, 19
Prader-Willi syndrome 37
Premature labor 95
Proteins, acidic 8
Pseudohermaphrodite 55
  female 55
  male 55
Pubic hair, scanty 50

R

Reciprocal translocation 38, 38f
Renal abnormalities 45
Replacement therapy 96
Rh incompatibility 93
Rh sensitization 95
Ribose nucleic acid 8, 60
Robertsonian translocation 38, 38f
Rocker bottom feet 46

S

Sex chromatin, study of 10
Sex chromosomes, disorders affecting 48
Sex-linked genes 57
Sex-linked inheritance 71, 75
Shoulders, hypermobility of 46
Sickle cell anemia 73, 80, 80t
Sickle cell trait 80, 80t
Skin fibroblast 11
Somatic cell 33
Squamous cell 17f
Stenosis, pulmonary 50

T

Testicular feminization 48, 77
    syndrome 54
Thalassemia 73
Transfer RNA 60
Triple X syndrome 17, 48, 53
Turner's syndrome 17, 19, 33, 35, 48, 49f, 83, 92

Twins
    conjoined 89
    dizygotic 86, 87, 88f
    monozygotic 86, 87f
    study of 86

U

Umbilical cord perforation 95
Urinary tract anomalies 50

V

Varicose veins 53
Viruses 35
Vision, defective 50
Vitamin D
    replacement of 96
    resistant rickets 78, 96

X

X-linked dominant inheritance 71, 76, 77
X-linked recessive inheritance 76t, 77t
    transmission 76f, 78f
XYY syndrome 48, 54

Y

Y-linked inheritance 71, 78

Z

Zygosity, tests of 89
Zygotene 29